CONTENTS

# Introduction

| | |
|---|---|
| Foreword | 2 |
| About this book | 3 |
| About the test | 3 |
| Preparing for a test | 7 |
| Booking a test | 9 |
| Taking a test | 11 |
| Augmented reality | 12 |

# INTRODUCTION

## Foreword

If you are reading this, you're probably getting ready to take the *Health, safety and environment test*. If you've not worked on site before, this test might seem a bit of a bind – what's the point? The answer is that when you work on site you are not just responsible for the health and safety of yourself but also of those working around you.

You might ask, 'Why is this – surely I can't be expected to look after others I don't even know?' The law says that you have a duty of care, but beyond any legal requirement, it's also the right thing to do. Even if you don't set people to work, you still have to look out for others. It's part of the job.

Construction is still one of the most dangerous industries to work in. You might be thinking that nothing is likely to go wrong on site, but each year around 80,000 workers suffer an illness that they believe was caused or made worse by their work. That's over 300 workers a day! Around 52,000 of these are cases of bad backs, damaged shoulders or similar injuries, 12,000 are due to stress, anxiety and depression and 3,000 are from breathing and lung problems. Although the fatal accident figures have generally improved over the last 20 years, construction workers are still dying from work-related causes. I'm sure we can all agree that even one death is one too many.

By now you might be thinking, 'OK, but what can I do? I'm just one person.' By studying this book and by taking CITB's *Health, safety and environment test*, you are making a great start. Whether you're new on site or an experienced worker, CITB's revision materials and test will help you get the basic health, safety and environmental knowledge and skills so that you can stay safe on site, spot dangers and speak up when you see that things might be going wrong.

CITB works with industry experts and construction workers to make sure that the *Health, safety and environment test* is up-to-date and fit for purpose. The test is always being reviewed and new question styles have been introduced.

We all want to make sites safe places where no-one's health is harmed. CITB is here to help you do that. Working together, looking out for each other, we can raise health and safety standards and make a positive impact on our industry.


Revd. Eur Ing Kevin Fear BSc(Hons), CEng, MICE, CMIHT, CMIOSH, Hon FaPS
Health and Safety Strategy Lead
CITB

# CONTENTS

**Introduction**

| | |
|---|---|
| Foreword | 2 |
| About this book | 3 |
| About the test | 3 |
| Preparing for a test | 7 |
| Booking a test | 9 |
| Taking a test | 11 |
| Augmented reality | 12 |

**A** **Working environment**

| | | |
|---|---|---|
| **01** | General responsibilities | 14 |
| **02** | Accident reporting and recording | 20 |
| **03** | First aid and emergency procedures | 24 |
| **04** | Personal protective equipment | 28 |
| **05** | Environmental awareness and waste control | 33 |

**B** **Occupational health**

| | | |
|---|---|---|
| **06** | Dust and fumes (Respiratory hazards) | 42 |
| **07** | Noise and vibration | 53 |
| **08** | Health and welfare | 59 |
| **09** | Manual handling | 72 |

**C** **Safety**

| | | |
|---|---|---|
| **10** | Safety signs | 78 |
| **11** | Fire prevention and control | 81 |
| **12** | Electrical safety, tools and equipment | 86 |

# CONTENTS

## D High risk activities

**13** Site transport and lifting operations     96
**14** Working at height     103
**15** Excavations and confined spaces     115
**16** Hazardous substances     119

## E Specialist

If you are preparing for a specialist test you also need to revise the appropriate specialist activity, from those listed below.

**17** Supervisory     126
**18** Demolition     140
**19** Highway works     147
**20** Specialist work at height     156
**21** Lifts and escalators     165
**22** Tunnelling     174

Heating, ventilation, air conditioning and refrigeration (HVACR)

**23** Heating and plumbing services     182
**24** Pipefitting and welding     190
**25** Ductwork     199
**26** Refrigeration and air conditioning     207
**27** Services and facilities maintenance     215

**28** Plumbing (JIB)     223

### Further information

Answer pages     232
Acknowledgements

## About this book

This book has been created to help you revise for your *Health, safety and environment test*. It contains all of the content that is covered within the test through questions and answers or statements of information so that you can fully prepare for your test.

The book also includes information about how to book your test, any special assistance available and other helpful topics.

## About the test

The CITB *Health, safety and environment test* helps raise standards across the industry. It ensures that workers meet a minimum level of health, safety and environmental awareness before going on site.

The test structure has been designed to enable you to demonstrate knowledge across the following key areas.

**Section A: Working environment**

**Section B: Occupational health**

**Section C: Safety**

**Section D: High risk activities**

**Section E: Specialist**

### Section A: Working environment

**General responsibilities:** what you and your employer need to do to ensure everyone is working safely on site.

**Accident reporting and recording:** when, how and why accidents need to be reported and recorded.

**First aid and emergency procedures:** what you should do in case of an emergency, and what your employer must make available.

**Personal protective equipment:** why personal protective equipment (PPE) is important, why you should wear it and who is responsible for it.

**Environmental awareness and waste control:** your responsibilities on site, how waste should be managed and how to conserve energy.

### Section B: Occupational health

**Dust and fumes (Respiratory hazards):** how to work safely, protecting yourself and those around you from exposure to respiratory hazards. What health conditions may arise from exposure to dust and fumes.

**Noise and vibration:** why it is important to minimise exposure to noise and vibration in the workplace. How you should protect yourself and those around you.

**Health and welfare:** common health issues on site and how to avoid them. Providing welfare facilities and support on site. Awareness of mental health.

**Manual handling:** why and how it is important to handle all loads using a safe system of work. What key areas you need to be aware of when handling loads.

# INTRODUCTION

**Safety signs:** what types of safety sign you will see on a construction site, and what they are informing you of.

**Fire prevention and control:** what you should do if you discover a fire, and which fire extinguishers should be used on what type of fire.

**Electrical safety, tools and equipment:** how to work safely with different types of tools, and what you should do if the tools you are using have not been examined or are faulty.

### Section D: High risk activities

**Site transport safety and lifting operations:** how careful planning can safely segregate pedestrian and traffic routes. Traffic rules you need to be aware of. How to lift loads safely.

**Working at height:** what types of equipment you will use for working at height, and how to use them correctly.

**Excavations and confined spaces:** how to work safely in a confined space or excavation, and what you should do if exposed to certain hazards.

**Hazardous substances:** how you can identify a hazardous substance, and what control measures should be in place to enable you to work safely.

### Section E: Specialist

If you are preparing for a specialist test you will also be asked questions about your specialist activity. There are currently 12 specialist tests available, including supervisory; demolition; highway works; specialist work at height; lifts and escalators; tunnelling; plumbing (JIB); heating and plumbing services; pipefitting and welding; ductwork; refrigeration and air conditioning; services and facilities maintenance.

## How is the test structured?

All tests last for 45 minutes and have 50 knowledge questions.

## What is a knowledge question?

The knowledge questions cover 16 core areas (presented in Sections A–D of this book) that are included in all the tests. These questions are factual. For example, they will ask you to identify fire extinguishers and signs. There is an additional knowledge question bank for each specialist test.

You do not need to have a detailed knowledge of the exact content of any regulations. However, you do need to show that you know what is required of you, the things you must do (or not do), and what to do in certain circumstances (for example, upon discovering an accident).

Legislation in Northern Ireland and Scotland differs from that in the rest of the UK. For practical reasons, all candidates (including those in Northern Ireland and Scotland) will be tested on questions using legislation relevant to the remainder of the UK only.

There are four different styles of knowledge question that may be presented within your test. These are explained below.

**Multiple choice and multiple choice with images**

 **Multiple-choice questions are identified by this icon.**

A multiple-choice question will ask you to select one or more answers from a list of options. Some answer options may also contain images.

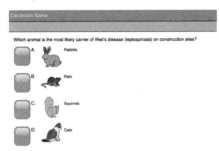

**Drag and drop text and drag and drop images**

 **Drag and drop questions are identified by this icon.**

A drag and drop question can be answered by dragging and dropping text or images from a list of options to the answer area.

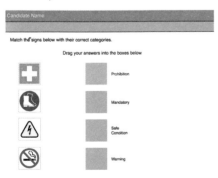

# INTRODUCTION

## Hot spot

 Hot spot questions are identified by this icon.

A hot spot question can be answered by selecting the correct place on the given image.

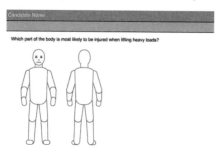

## Hot area

 Hot area questions are identified by this icon.

A hot area question can be answered by selecting one of the answer areas within the given image.

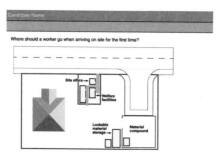

 **To practise these question styles online, visit www.citb.co.uk/hsandetest**

### Who writes the questions?

The question bank is developed by industry-recognised organisations and question writing experts alongside the health, safety and environment test question sub-committee.

We also work closely with industry to ensure the content covered in the test is relevant and fit for purpose.

### Will the questions change?

Health, safety and environment legislation, regulations and best practice will change from time to time, but CITB makes every effort to keep the test and the revision material up-to-date.

● You will not be tested on questions that are deemed to be no longer current.

● You will be tested on knowledge questions presented in the most up-to-date edition of the book. To revise effectively for the test you should use the latest edition. You can check which edition of the book you have at www.citb.co.uk/hsandetest or phone 0344 994 4488.

The *Health, safety and environment test* is currently under development. To find further information about the development taking place and how this affects you, visit www.citb.co.uk/hsandetest

## Preparing for a test

To pass your *Health, safety and environment test* you need to demonstrate knowledge and understanding across a number of areas, all of which are relevant to people working in a construction environment. The test is broken into sections so that knowledge across all key areas can be demonstrated.

There are a number of ways you can prepare for your test and increase your success.

# INTRODUCTION

| Revision material | Operatives | Specialists | Supervisors | Managers |
|---|---|---|---|---|
| ☺ Watch *Setting out* | This video will help you prepare for the behavioural elements that are embedded in the test questions. | | | |
| ◻ Read the revision books | *HS&E test for operatives and specialists (GT100)* | | | *HS&E test for managers and professionals – (GT200)* |
| 📀 Use the digital products | *HS&E test for operatives and specialists* – DVD (GT100 DVD) – Download (GT100 DL) – App | | | *HS&E test for managers and professionals* – DVD (GT200 DVD) – Download (GT200 DL) – App |
| ◻ Read supporting knowledge material | *Construction health and safety awareness* (GE707) | *Construction health and safety awareness* (GE707) plus sector recommended supporting material | *Construction site supervision* (GE706) | *Construction site safety – The comprehensive guide* (GE700) |
| Complete an appropriate training course | Site Safety Plus – one-day *Health and safety awareness* course | Contact your industry body for guidance | Site Safety Plus – two-day *Site supervision safety training scheme* | Site Safety Plus – five-day *Site management safety training scheme* |

## Where can I buy additional revision material?

CITB has developed a range of revision material, including this GT100 book, DVDs, downloads and a smartphone app that will help you to prepare for the test. For further information and to buy these products:

 **www.citb.co.uk/hsanderevision**

 **0344 994 4488**

visit the highstreet or online for books and DVDs. Visit the Apple App store or the Google Play store for smartphone apps.

 for further products and services that CITB offers visit citb.co.uk

### What's on the DVD, app and download?

The DVD, app and download offer an interactive package that includes:

● combination of knowledge questions and answers and statements of information

● mock testing

● voiceovers in the 14 supported languages

● the *Setting out* film.

Please note: British Sign Language assistance is included on the GT100 revision DVD and within the *Setting out* film.

## Booking a test

The easiest way to book your test is either online or by telephone. You should be able to book a test at your preferred location within two weeks. You will be given the date and time of your test immediately and offered the opportunity to buy revision material (for example, a book, DVD, download or app).

To book your test:

 www.citb.co.uk/hsandetest

 0344 994 4488
Welsh booking line 0344 994 4490

 post in an application form (application forms are available from the website and the telephone numbers listed above).

When booking your test you will be able to choose whether to receive confirmation by email or by letter. It is important that you check the details (including the type of test, the location, the date and time and what ID is required at the test centre) and follow any instructions given regarding the test.

If you do not receive a confirmation email or letter within the time specified, please call the booking line to check your booking has been made.

# INTRODUCTION

## What information do I need to book a test?

To book a test you should have the following information to hand.

- Which test you need to take.

- Whether you require any special assistance (see below).

- Your chosen method of payment (debit or credit card details).

- Your personal details.

- Your CITB registration number, if you have taken a *Health, safety and environment test* before or applied for certain card schemes.

## What special assistance is available when taking the test?

### Operatives

- You can book an operatives test with a voiceover in the following languages: Bulgarian, Czech, English, French, German, Hungarian, Lithuanian, Polish, Portuguese, Punjabi, Romanian, Russian, Spanish and Welsh when booking your test online or over the phone.

### Specialist tests

These can be booked with an English or Welsh voiceover when booking online or over the phone, or alternatively you can book an interpreter for any other language by calling the special assistance number below.

### Further assistance

- The test for operatives can be booked with British Sign Language on screen. If you need assistance in the other tests a signer can be provided.

- Other special assistance available includes a reader, reader recorder, signer, or extra time. These and the other special assistance mentioned above can be booked via a dedicated booking line: 0344 994 4491.

## How do I cancel or reschedule my test?

To cancel or reschedule your test you should go online or call the booking number at least 72 hours (three working days) before your test. There will be no charge for cancelling or rescheduling the test online at www.citb.co.uk/hsandetest outside of the 72-hour period. Reschedules and cancellations made via the telephone booking line will incur an administration fee.

## Taking a test

**On the day of the test you will need to:**

- allow plenty of time to get to the test centre and arrive at least 15 minutes before the start of the test

- take your confirmation email or letter

- take proof of identity that includes your photo and your signature (such as a driving licence card or passport – a full list of these requirements can be found on your booking confirmation or online at www.citb.co.uk/hsandetest

On arrival at the test centre, staff will check your documents to ensure you are booked onto the correct test. If you do not have all the relevant documents you will not be able to sit your test and you will lose your fee.

**During the test**

The tests are all delivered on a computer screen. However, you do not need to be familiar with computers and the test does not involve any typing. All you need to do is select the relevant answer(s), using either a mouse or by touching the screen.

Before the test begins you will work through a tutorial. It explains how the test works and lets you try out the buttons and functions that you will use while taking your test.

There will be information displayed on the screen which shows you how far you are through the test and how much time you have remaining.

**After the test**

At the end of the test there is an optional survey that gives you the chance to provide feedback on the test process.

You will be provided with a printed score report after you have left the test room. This will tell you whether you have passed or failed your test, and give feedback on areas where further learning and revision are recommended.

### What do I do if I fail?

If you fail your test, your score report will provide you with information on the areas where you got questions wrong.

It is strongly recommended that you revise these areas thoroughly before re-booking. You will have to wait at least 48 hours before you can take the test again.

### What do I do if I pass?

Once you have passed your test, you should consider applying to join the relevant card scheme, if you have not done so already. However, please be aware that you may need to complete further training, assessment and/or testing to meet their specific entry requirements.

Your pass certificate will also include any areas of the test in which you answered questions incorrectly. It is important that you improve your knowledge in these areas.

# INTRODUCTION

To find out more about many of the recognised schemes:

 **www.citb.co.uk/cards-testing/**

## Fraudulent testing

If you are aware of any fraudulent activity in the delivery of your test, or relating to cards or training in the construction industry:

 **email our fraud investigation team at report.it@citb.co.uk**

CITB takes reports of fraud linked to our testing processes extremely seriously. Working with the Police and other law enforcement agencies, we are doing everything we can to address the issue. Where possible, we always prosecute those engaged in any fraudulent activity.

## Augmented reality

### What is augmented reality?

Augmented reality (AR) technology is used in this revision book to provide readers with additional digital content such as videos, images and weblinks.

This technology has been used to connect you with extra content, free of charge. It can be accessed via your mobile device using the Layar app.

 **How to install and use the Layar app**

- Go to the appropriate app store (Apple or Android) and download the Layar app (free of charge) to your mobile device.

- Look out for the AR logo (right), which indicates a Layar-friendly page or image.

- Open the Layar app and scan the Layar-friendly page or image (ensure that you have the whole page or the specific image in view whilst scanning).

- Wait for the page to activate on your device.

- Select one of the buttons, when they appear, to access the additional content.

### Where can I find augmented reality in this revision book?

The following content is accessible by scanning the cover page of this publication.

- Book a *Health, safety and environment test*.

- Watch the *Setting out* film.

- Buy additional revision material.

## CONTENTS

# Working environment

**01**    General responsibilities    14

**02**    Accident reporting and recording    20

**03**    First aid and emergency procedures    24

**04**    Personal protective equipment    28

**05**    Environmental awareness and waste control    33

# WORKING ENVIRONMENT

## 01    General responsibilities

- The Health and Safety at Work etc. Act 1974 contains legal duties for employers and employees.

- Visitors and workers must be given a site induction and authorised to enter site.

- Employers must provide information to workers about site rules, welfare facilities and emergency procedures.

- Workers should be provided with clean welfare facilities and information regarding hazards and risks at work.

- Everyone on site is responsible for the consideration of neighbours and members of the public.

- Employees should follow a safe system of work agreed with the employer.

- A safe system of work would include information such as the sequence of work, and any hazards associated with the task.

- The purpose of a risk assessment is to identify hazards and control risk.

- Risk assessments consider the likelihood of a hazard occurring and the seriousness of harm that could occur.

- A method statement will inform a worker of the safe way to carry out a task.

- If a task feels dangerous or unsafe, stop work and report it immediately.

---

**1.01**  The whole site has been issued with a prohibition notice. What does this mean?

- A  Continue with site work

- B  Finish the job and go home

- C  Do not use any power tools

- D  Stop work because the site is unsafe

**1.02**  After watching you work, a Health and Safety Executive (HSE) inspector issues an improvement notice. What does this mean?

- A  You are not working fast enough

- B  You are not working in a safe way

- C  Your work has improved since the last visit

- D  You need to improve the standard of your work

**1.03** You have witnessed a serious accident on your site and are to be interviewed by a Health and Safety Executive (HSE) inspector. What **should** you do?

A Ask other workers what you should tell the inspector

B Ask your supervisor what you should tell the inspector

C Co-operate and tell the inspector exactly what you saw

D Not tell the inspector anything, and ask them to talk to your supervisor

**1.04** If you notice that a design detail can't be built in the way it has been drawn in the plans, what **two** things should you do?

You will be asked to 'drag and drop' your answers

A Leave that detail out altogether

B Build it as you think it should be done

C Keep quiet as it will mean more work for you

D Only make the changes when they are approved in writing

E Raise the issue with your supervisor before you start work

**1.05** When workers arrive on site what is the **first** thing they **should** do?

A Walk around the site to inspect the work from the day before

B Enter the site by the easiest route and start work

C Get their tools out of the store and start work

D Make sure that the site team knows they are there

**1.06** If a worker fails to report a near miss, what **could** happen?

A The company could go out of business through neglect

B The employee could get a large fine

C The near miss could be a serious accident next time

D The site manager will be sacked immediately

**1.07** What are **two** possible consequences for you if your employer does **not** prevent accidents and ill health at work?

You will be asked to 'drag and drop' your answers

- A  You will have to work longer hours to earn more money
- B  You may suffer an injury, affecting your health and wellbeing
- C  You won't get the training required to continue working on site
- D  You may not be able to work, which would affect your income and family life
- E  You will have worse welfare facilities on site while improvements are made

**1.08** What are **two** possible consequences for employers of **not** taking measures to prevent accidents and ill health at work?

You will be asked to 'drag and drop' your answers

- A  They could be fined or imprisoned
- B  They will damage the environment
- C  They will need to employ more people
- D  They will have to change the site layout for emergency vehicles
- E  They will lose time and money due to the cost of any accident or ill health

**1.09** What does the word **hazard** mean?

- A  Anything that could cause harm
- B  The construction site accident rate
- C  The likelihood of something happening
- D  A type of removable barrier or machine guard

**1.10** What is the **main** reason for understanding the fire and emergency procedures on site?

- A  To know where the fire exits and assembly points are in an emergency
- B  To know what tools and equipment can be used during an emergency
- C  To help you to get time off work in an emergency
- D  To stop anyone leaving site in an emergency

**1.11** Who is responsible for managing health and safety on site?

A. Site manager

B. Building inspector

C. Contracts manager

D. Health and Safety Executive (HSE)

**1.12** Why is it the employer's legal responsibility to discuss matters of health and safety with employees?

A. So that employees do not have any responsibilities for health and safety

B. So that employees will never have to attend any other health and safety training

C. So that employees are informed of things that will protect their health and safety

D. So that your employer will not have any legal responsibility for employees' health and safety

**1.13** General site rules would **not** normally include information about which one of the following?

A. Personal protective equipment (PPE)

B. Names and addresses of workers

C. Near miss and accident reporting

D. Site induction procedures

**1.14** If someone is injured on site, where **should** this be recorded?

A. In an accident book or record

B. On the safe system of work plan

C. On the site plan

D. In the method of work

**1.15** What should all risk assessments identify?

A. The site working hours

B. How to report accidents

C. Where the first-aid kit is kept

D. The hazards in the work environment

**1.16** When creating a risk assessment the severity of harm is multiplied by what?

A. The number of workers on site

B. The likelihood of harm occurring

C. The cost of injury or harm

D. The area of the construction site

**01**

**1.17** Which **two** topics should be covered in a site induction?

You will be asked to 'drag and drop' your answers

- A | Site rules
- B | Local amenities
- C | Holiday entitlement
- D | Local transportation links
- E | Site emergency procedures

**1.18** How would you expect to find out about health and safety rules when you **first** arrive on site?

- A | During the induction
- B | In a letter sent to your home
- C | By reading the health and safety policy
- D | By asking other workers to show you around

**1.19** What is a toolbox talk?

- A | A sales talk given by a tool supplier
- B | A talk that tells you where to buy tools
- C | Your first training session when you arrive on site
- D | A short training session on a particular safety topic

**1.20** What is the **main** reason for attending a site induction?

- A | To get to know other new employees
- B | Site rules and hazards will be explained
- C | To create the method statements for the site
- D | Permits to work will be written and handed out

**1.21** What **should** you do if the safety rules given in your site induction seem out of date as work progresses?

- A | Speak to your supervisor about your concerns
- B | Nothing, as safety is the site manager's responsibility
- C | Speak to your workmates to see if they have any new rules
- D | Make up your own safety rules to suit the changing conditions

**1.22** During the site induction you do **not** understand something the presenter says. What **should** you do?

A Attend another site induction

B Ask the presenter to explain it again

C Guess what the presenter was saying

D Wait until the end, then ask someone else to explain

**1.23** Employers must provide workers with instructions that meet which requirement?

A Downloadable from the internet

B Written in large print

C Available in audio

D In a format each worker understands

**1.24** A worker finds a way of working that is quicker than the method statement they have been given. What **should** they do?

A Inform work colleagues so they can work this way

B Get their work done more quickly so they can leave early

C Get more work done so they can earn more money

D Continue to follow the safe system of work for the task

**1.25** Who **should** you speak to if the work of another contractor is affecting your safety?

A Your supervisor

B The contractor

C Your workmates

D The contractor's supervisor

**1.26** What **should** you do if you **cannot** do a job in the way described in the method statement?

A Make up a better way to do it and carry on

B Contact the Health and Safety Executive (HSE)

C Ask other workers how they think it should be done

D Do not start work until you have talked to your supervisor

**1.27** What **should** a worker do if the helmet they are using is damaged?

A Use it but keep checking it

B Put a sticker over the damaged area

C Report it at the end of the day

D Replace it immediately

**02**

**02    Accident reporting and recording**

- Reporting unsafe conditions is everyone's responsibility on site.
- Reporting near misses will help to prevent them happening again.
- Plant and machinery should only be used by authorised and competent operatives.
- Any accident causing injury must be recorded in an accident book.
- All relevant staff should be involved in investigating accidents and near misses.

**2.01** You suffer an injury at work and the details are recorded in the accident book. What **must** happen to this accident record?

| A | It must be kept in a place where anyone at work can read it |
| B | It must be sent to the insurance company at the end of the job |
| C | It must be treated as confidential under data protection laws |
| D | It must be destroyed at the end of the job, due to confidentiality |

**2.02** What **must** be done if an operator is driving plant equipment faster than site speed limits?

| A | Alert all other staff on site to be careful |
| B | Inform a supervisor or manager |
| C | Shout at the driver, telling them to slow down |
| D | Wait until they stop and talk to them about it |

**2.03** In order to reduce the risk of accidents, which **one** of the following **should** be avoided when driving vehicles on site?

| A | Use designated turning areas |
| B | Implement a one-way system around the site |
| C | Drive-through loading and unloading areas |
| D | Reverse without the use of a vehicle marshaller |

**2.04** Which **two** of the following would result in you being **ordered** off site?

| A | Losing your road users' driving licence |
| B | Being under the influence of alcohol |
| C | Driving downhill with a heavy load |
| D | Driving without using the flashing beacon |
| E | Being under the influence of drugs |

**2.05** You have been injured in an accident at work and, as a result, are absent for more than seven days. Which **two** of the following actions **must** be taken?

You will be asked to 'drag and drop' your answers

| | |
|---|---|
| A | The accident must be recorded in the accident book |
| B | The local hospital and the benefits office must be informed |
| C | You must pay for any first-aid equipment used to treat your injury |
| D | Your employer must inform the Health and Safety Executive (HSE) |
| E | The emergency services must be called to find out how the accident happened |

**2.06** If you have a minor accident, who **should** report it?

A — You, if possible

B — The sub-contractor

C — Anyone who saw the accident

D — The Health and Safety Executive (HSE)

**2.07** You are injured in an accident at work. When **should** you report it?

A — The next day before you start work

B — Immediately, or as soon as possible

C — Only if you have to take time off work

D — At the end of the day, before you go home

**2.08** Why **should** you report an accident?

A — It is a legal requirement

B — It helps the site find out who caused it

C — So that everyone can find out what happened

D — So that your company will be held responsible

**2.09** Who **must** you report a serious accident to?

A — Site security

B — Your employer

C — The police service

D — The ambulance service

# WORKING ENVIRONMENT

02

**2.10** What action should be taken if you witness a serious accident on site?

A. Telephone the local doctor for advice

B. Tell your supervisor that you saw what happened

C. Say nothing in case you get someone into trouble

D. Ask your workmates what they think you should do

**2.11** Which of the following statements **best** describes a near miss?

A. An incident that nearly resulted in injury or damage

B. An incident where you were just too late to see what happened

C. An incident where someone was injured and nearly had to go to hospital

D. An incident where someone was injured and nearly had to take time off work

**2.12** While working on site you cut one of your fingers. What should you do?

A. Report it and get first aid if necessary

B. Clean it and tell your supervisor about it later

C. Wash it, and if it is not a problem carry on working

D. Report it at the end of the day or the end of the shift

**2.13** What is the **main** objective of carrying out an accident investigation?

A. To place blame

B. To identify the people involved

C. To find the cause and prevent recurrence

D. To help track the cost of insurance claims

**2.14** A scaffold has collapsed and you saw it happen. What should you say when you are asked about the accident?

A. Who you think should be blamed and punished

B. Exactly what you saw, giving as much detail as possible

C. As little as possible because you are not a scaffold expert

D. As little as possible because you don't want to get people into trouble

**2.15** Which **two** of the following are the **main** reasons for reporting accidents, incidents and near misses?

A. To find out whom claims should be made against

B. To understand how and why things went wrong

C. Certain incidents or accidents have to be reported to the Health and Safety Executive (HSE)

D. To make sure none of the supervisors find out about the accident

E. To help the company avoid being prosecuted or fined

**2.16** Which **two** of the following items should be recorded in the accident book?

You will be asked to 'drag and drop' your answers

| A | Injuries sustained |
|---|---|
| B | Date of the accident |
| C | Telephone number |
| D | Location of the hospital |
| E | National insurance number |

**2.17** If someone is injured at work, who **should** record it in the accident book?

A  The first aider identified on site

B  The company contract manager

C  The injured person or someone acting for them

D  Someone from the Health and Safety Executive (HSE)

**2.18** Which of the following does **not** have to be recorded in the accident book?

A  Details of the injury sustained

B  The injured person's home address

C  The date and time that the injury happened

D  The injured person's national insurance number

**2.19** Which of the following is the **least** important reason for recording all accidents?

A  It might stop them happening again

B  Details have to be entered in the accident book

C  To find out who is to blame and make sure they are prosecuted

D  Some accidents have to be reported to the Health and Safety Executive (HSE)

## 03    First aid and emergency procedures

- All first aiders should have a current, up-to-date first aid at work certificate.
- The place to go in the event of an emergency is called an **assembly point**.
- The location of the emergency assembly point should be identified in a site induction.

*Emergency
assembly point*

**3.01**    What should be done in the event of an emergency on site?

A    Follow the site emergency procedure

B    Collect your personal items and leave the site

C    Leave the site by the nearest exit and return home

D    Phone the Health and Safety Executive (HSE) for advice

**3.02**    Which **two** of the following will help you to find out about the site emergency procedures and emergency telephone numbers?

A    Attending the site induction

B    Reading the site noticeboards

C    Looking in the telephone directory

D    Guidance from your local job centre

E    Guidance from the Health and Safety Executive (HSE) website

**3.03**    How **should** you be informed about what to do in an emergency? Give **two** answers.

A    By asking at the local hospital

B    By attending the site induction

C    By reading the site noticeboards

D    By looking in the health and safety file

E    By asking the Health and Safety Executive (HSE)

**3.04** What **two** things should you do if there is an emergency situation on site?

You will be asked to 'drag and drop' your answers

| A | Finish what you are doing |
| B | Leave the area via the nearest exit |
| C | Go to the designated assembly point |
| D | Collect personal items from the site office |
| E | Look for other people who may not know what to do |

**3.05** What information **should** be gathered after a near miss incident occurs?

A | The names of next of kin for the people involved

B | Where those involved lived at the time of the incident

C | The activities that were being carried out at the time

D | The cost of the project at the time of the incident

**3.06** You witness a serious accident on site. What **immediate** action should you take? Give **two** answers

A | Call out to other workers so they can call for help

B | Check if it is safe to approach the injured person

C | Sit the injured person up and give them food and water

D | Record the date and time in the incident book

E | Lift the injured person and take them to the site office

**3.07** What should **not** be in a first-aid kit?

A |  Bandages

B | 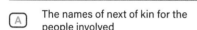 Plasters

C |  Safety pins

D |  Tablets and medicines

**3.08** Does your employer have to provide a first-aid kit?

A | Yes, every site must have one

B | Only if more than 25 people work on site

C | Only if more than 50 people work on site

D | No, there is no legal duty to provide one

**3.09** If the first-aid kit on site is empty, what **should** you do?

A Bring your own first-aid supplies into work

B Ignore the problem as it is always the same

C Find out who is taking all the first-aid supplies

D Inform the person who looks after the first-aid kit

**3.10** What is the one thing a first aider **cannot** do?

A Stop any bleeding

B Treat you if you are unconscious

C Give mouth-to-mouth resuscitation

D Give you medicines without authorisation

**3.11** Evacuation routes **should** be:

A lit at all times of the day

B painted bright green

C used as assembly points

D clear and unobstructed

**3.12** If you find an injured person and you are on your own, what should you do **first**?

A Assess the situation – do not put yourself in danger

B Inform your supervisor that someone has been injured

C Move the injured person to a safe place, and then find your supervisor

D Ask the injured person what happened, and then find your supervisor

**3.13** Someone working in a deep inspection chamber has collapsed. What should you do **first**?

A Climb into the inspection chamber and give first-aid treatment

B Get someone to lower you into the inspection chamber on a rope

C Raise the alarm and stay by the inspection chamber, but do not enter

D Ask someone to find your supervisor while you try to rescue the worker

**3.14** Someone is knocked unconscious and you are **not** trained in first aid. What should you do **first**?

A Send for medical help

B Slap their face to wake them up

C Give them mouth-to-mouth resuscitation

D Turn them over so that they are lying on their back

**3.15** Someone has fallen from height and has no feeling in their legs. What **should** you do?

A Keep them still until medical help arrives

B Roll them onto their side and bend their legs

C Raise their legs to see if any feeling comes back

D Keep their legs straight and roll them onto their back

**3.16** Someone collapses with stomach pain and there is no first aider on site. What should you do **first**?

A Ask them to sit down

B Get them to take some painkillers

C Ask someone to call the emergency services

D Help them to lie down in the recovery position

**3.17** If you think someone has broken their leg, what **should** you do?

A Place them on their back

B Send for the first aider or get other help

C Use your belt to strap their legs together

D Place them on their side in the recovery position

**3.18** If you cut your finger and it won't stop bleeding, what **should** you do?

A Wash it, then carry on working

B Find a first aider or get other medical help

C Wrap something around it and carry on working

D Tell your colleagues because you may need to rest

**3.19** If there is an emergency while you are on site, what should you do **first**?

A Leave the site and go home

B Phone home and then leave the site

C Follow the site emergency procedure

D Phone the Health and Safety Executive (HSE)

**3.20** If someone is in contact with a live cable, what should you do **first**?

A Phone the electricity company

B Pull them away from the cable

C Isolate the power and call for help

D Dial 999 and ask for an ambulance

# WORKING ENVIRONMENT

## 04 Personal protective equipment

- Employers should provide workers with personal protective equipment (PPE) and the means to maintain it correctly, free of charge.
- Wearing PPE will help to protect workers from physical injury or ill health.
- Size and fit should be considered in the selection of suitable PPE.
- Stop work immediately and replace PPE if it gets damaged.

**4.01** When **must** your employer provide you with personal protective equipment (PPE)?

- A  Twice a year
- B  If you pay for it
- C  If it is in the contract
- D  If you need to be protected

**4.02** If you have to work outdoors in bad weather, why **should** your employer supply you with waterproof clothing?

- A  To keep you warm and dry, so you take fewer breaks
- B  To protect you from the weather, which will reduce trips and falls
- C  To keep you warm and dry, so you are less likely to catch Weil's disease (leptospirosis)
- D  To protect you from the weather, as you are less likely to get muscle strains if you are warm and dry

**4.03** Which of the following statements about personal protective equipment (PPE) is **not** true?

- A  You must use it as instructed
- B  You must pay for any damage or loss
- C  You must store it correctly when you are not using it
- D  You must report any damage or loss to your supervisor

**4.04** Which of the following statements about wearing a safety helmet in hot weather is **true**?

- A  You can modify it to keep your head cool
- B  You must wear it at all times and in the right way
- C  You must take it off during the hottest part of the day
- D  You can wear it back-to-front if it is more comfortable that way

**4.05** What should you wear if there is a risk of materials flying into your eyes?

A. Tinted welding goggles

B. Laser safety glasses

C. Chemical-rated goggles

D. Impact-rated goggles

**4.06** When using a grinder or cut-off saw, what type of eye protection should be worn?

A. Impact-rated goggles or full face shield

B. Light eye protection (safety glasses)

C. Reading glasses or sunglasses

D. Welding goggles

**4.07** When **should** you wear safety footwear on site?

A. All the time

B. Only when working inside

C. Until the site starts to look finished

D. Only when working at ground level

**4.08** When is the only time that you do **not** need to wear head protection on site?

A. If you are self-employed

B. If you are working alone

C. If you are in a safe area, like the site office

D. If you are working in hot weather

**4.09** When you start a new task, how will you know if you need any **extra** personal protective equipment (PPE)?

A. You will always need it

B. By looking at the risk assessment

C. By looking at the company webpage

D. By looking at your employer's health and safety policy

**4.10** What is the **main** risk to this worker, wearing **only** these items of personal protective equipment (PPE)?

| A | Dermatitis to skin |
|---|---|
| B | Damage to hearing |
| C | Eye injuries |
| D | Breathing in harmful dust |

**4.11** When selecting appropriate personal protective equipment (PPE), what is the **most** important factor to be taken into account?

| A | The type of hazard |
|---|---|
| B | Can it be recycled |
| C | The cost of the equipment |
| D | How long it will last |

**4.13** What will safety footwear with a protective mid-sole protect you from?

| A | Spillages, which may burn the sole of your foot |
|---|---|
| B | Blisters, which could occur in warm, wet conditions |
| C | Twisting your ankle, as they have better grip than regular shoes |
| D | Nails or sharp objects, which could puncture the sole of your foot |

**4.12** What additional measures can be worn under a hard hat in cold weather?

| A | A baseball cap |
|---|---|
| B | A jumper with a detachable hood |
| C | A woolly hat |
| D | A manufacturer's attachment |

**4.14** What condition **could** be prevented if the correct gloves are worn while handling a hazardous substance?

| A | Arthritis |
|---|---|
| B | Skin disease |
| C | Vibration white finger |
| D | Raynaud's syndrome |

**4.15** Will all types of glove protect your hands against chemicals?

A. Yes, all gloves are made to the same standard

B. Only if you cover the gloves with barrier cream

C. Only if you put barrier cream on your hands first

D. No, different gloves protect against different types of hazard

**4.16** Which item of personal protective equipment (PPE) is helping to protect the worker from dermatitis?

**4.17** Good quality personal protective equipment (PPE) will be marked with which letter or letters?

A.  $C \epsilon$

B.  ©

C.  HSE

D.  ®

**4.18** How **should** a safety helmet be worn to get maximum protection from it?

A.  Back to front

B.  Pushed back on your head

C.  Square on your head

D. Pulled forward

**4.19** What should you do if your disposable, foam earplugs keep falling out?

A. Throw them away and work without them

B. Put rolled-up tissue paper in each ear instead

C. Put two earplugs in each ear so that they stay in place

D. Stop work until you are shown how to fit them properly

**4.20** If you need to wear a full body harness and you have **not** used one before, what **should** you do?

A. Try to work it out for yourself

B. Ask for expert advice and training

C. Read the manufacturer's instruction book

D. Ask someone wearing a harness to show you what to do

**04**

**4.21** Which of the following figures is wearing the **correct** personal protective equipment (PPE)?

A

C

B

D

**4.22** Where on the body would a worker wear respiratory protective equipment (RPE)?

## 05 Environmental awareness and waste control

- Everyone is responsible for minimising the amount of waste generated on site.

- Following the site environmental risk assessment will help to prevent pollution on a construction site.

- Segregating waste materials supports recycling and helps to avoid pollution.

*Waste skips*

- Re-using leftover materials helps to save energy and conserves raw materials.

- Recycling construction materials avoids waste going to landfill.

- Everyone on site should take responsibility for saving energy and water by turning off plant, equipment and taps when not in use.

- A good way of reducing energy if heating or cooling systems are being used in site accommodation is to keep windows and doors closed.

- Spill kits should be available to clean up spilt chemicals and oils.

- Bats and badgers are classed as protected species, and are protected by law.

- Many historic buildings are listed and protected by law; permission is required before making any changes to them.

**5.01** What **should** be done with waste concrete and washout water?

A. Bury it on site, as it will break down over time

B. Pour it down a drain with plenty of water

C. Bury it in a disposable bin liner

D. Place it in a lined skip for recycling

**5.02** This label is shown on the container of a liquid that a worker is using on site. What does it mean?

A. It can be used to feed plants and fish

B. It is harmful to the environment

C. It could cause a drought

D. It will only cause death to fish

**05**

**5.03** Which **two** of the following are common causes of water pollution on sites?

You will be asked to 'drag and drop' your answers

**A** Fuels being stored incorrectly and too close to drains

**B** Rain water washing material out of skips into surface water drains

**C** Exhaust gases from mobile plant getting into drainage systems

**D** Smoking and e-smoking near drainage systems

**E** Walkways freezing in winter near drainage systems

**5.04** Which one of the following is **true** of a spill on site, involving just **one** litre of oil?

**A** It is too small to cause a problem

**B** The main problem is that oil is expensive

**C** It will contaminate the ground

**D** It could cause serious air pollution

**5.06** What are **two** of the **best** ways of helping to save energy on site and reduce harmful emissions?

**A** Switch off plant and equipment, including generators, when they are not in use

**B** Keep windows and doors closed in offices and welfare facilities when the heating is on

**C** Report any defective, non-powered hand tools so that they can be repaired or replaced

**D** Use a generator rather than mains electricity for the offices and small items of equipment

**5.05** Which **three** statements are reasons why saving energy is important?

You will be asked to 'drag and drop' your answers

**A** It helps to reduce fuel and energy bills on site

**B** It helps to increase energy use on other sites

**C** It helps to save natural resources used to generate energy

**D** It helps energy companies to charge more for their services

**E** It helps to reduce the impact of climate change caused by burning fossil fuels

**5.07** Over ordering materials can result in what?

| A | Accidents |
|---|---|

| B | Waste |
|---|---|

| C | Danger |
|---|---|

| D | Lower costs |
|---|---|

**5.08** You are on site and need to throw away some waste liquid that has oil in it. What should you do?

| A | Pour it down a drain or sink in the welfare facilities |
|---|---|

| B | Pour it slowly onto the ground and let it soak away |
|---|---|

| C | Pour it into a sealed container and put it into a general waste skip |
|---|---|

| D | Ask your supervisor what the disposal process is for contaminated liquid |
|---|---|

**5.09** Which **two** items are classed as hazardous waste?

| A | Broken bricks |
|---|---|

| B | Untreated timber off-cuts |
|---|---|

| C | Panes of glass |
|---|---|

| D | Fluorescent light tubes |
|---|---|

| E | Used spill kits |
|---|---|

**5.10** Which items are hazardous waste and which are non-hazardous waste?

Non-hazardous

| A | Fluorescent light tubes |
|---|---|

| B | Broken bricks |
|---|---|

| C | Untreated timber off-cuts |
|---|---|

| D | Oil-based paint |
|---|---|

Hazardous

# WORKING ENVIRONMENT

**5.11** What is the **correct** way to clean up oil that has leaked from machinery onto the ground?

A) Put the oily soil into the general waste skip

B) Wash the oil away with water and detergent

C) Mix the soil up with other soil so that the oil cannot be seen

D) Put the oily soil into a separate container for collection as hazardous waste

**5.14** What should be done if there is an oil or diesel spill on site?

A) Use a spill kit to clean it up before the end of the day

B) Ignore it. Oil or diesel spills do not have serious, long-term effects

C) Stop work, contain the spill, notify the supervisor and then clean up the spill

D) Call the Environment agency immediately so they can arrange to have it cleaned up

**5.12** How should hazardous waste be dealt with on site? Give **two** answers.

You will be asked to 'drag and drop' your answers

A) Put it in a mixed waste skip

B) Segregate it from other waste

C) It can be put in any skip on site

D) Place it in the correctly labelled container

E) Take it to the nearest Local Authority waste tip

**5.13** Under environmental law, which of the following statements is **true**?

A) Only directors can be prosecuted if they do not follow the law

B) Only companies can be prosecuted if they do not follow the law

C) Only employees can be prosecuted if they do not follow the law

D) Companies and employees can be prosecuted if they do not follow the law

**5.15** Which of the following is **most** likely to cause air pollution?

A) Fuel spillage

B) Using diesel engines

C) Surface run-off

D) Excessive noise

**5.16** Which **two** actions could help minimise waste?

You will be asked to 'drag and drop' your answers

| | |
|---|---|
| A | Use new materials at the beginning of each day |
| B | Always take much more than required, just in case you need it |
| C | Leave bags of cement and plaster out in the rain, unprotected |
| D | Only take or open what you need and return or reseal anything left over |
| E | Reuse off-cuts (such as half bricks) rather than discarding them |

**5.17** What are the **two** most important reasons why waste should be segregated on site?

You will be asked to 'drag and drop' your answers

| | |
|---|---|
| A | The waste will take up less room in a skip |
| B | It is generally more cost effective to dispose of segregated waste |
| C | So that the wastes can be used or recycled more easily |
| D | So that the client can check what is being thrown away |
| E | To make sure that the labourer has enough work to do |

**5.18** Which of the following is **bad** practice?

A Storing materials safely

B Mixing all waste in one skip

C Refuelling carefully to avoid spills

D Switching off plant and equipment when it is not in use

**5.19** On site, waste **should** be collected in what?

A Segregated skips

B Bins and bays

C General skips

D Bays and buckets

**5.20** You discover a bird on a nest where you need to work. What should you do?

| A | Scare it away by making loud noises, then carry on with your work |

| B | Cover it with a sheet so it can be moved out of the way before starting work |

| C | Move it to a place of safety, carry out your work and then put it back |

| D | Protect it from further disturbance, make others aware and inform your supervisor |

**5.23** During excavation work, some interesting old coins are found in the loosened soil. What is the **most** appropriate action?

| A | Stop excavating the site and contact the supervisor |

| B | Keep excavating and see how many more there are to find |

| C | Keep quiet. The person who found them should keep them |

| D | Hide them. Archaeologists working on site will delay the works |

**5.21** Certain species of plants and animals in England are protected by law. A worker is breaking the law if they do which **two** things to the plant or animal?

You will be asked to 'drag and drop' your answers

| A | Report it |

| B | Photograph it |

| C | Remove it |

| D | Feed it |

| E | Destroy its habitat |

**5.22** Which of the following is an effective way to avoid causing harm to protected species?

| A | Only working at night |

| B | Avoiding breeding season |

| C | Take them to the site office |

| D | Using manually operated machinery |

**5.24** Preserving old buildings is important for contributing to an area's what?

| A | Historical record |

| B | Cost of living |

| C | Infrastucture |

| D | House prices |

**5.25** Which of the following does **not** cause a nuisance to neighbours of a building site?

A. Dust and fumes from the site

B. Carefully directed site lighting

C. Lorries and heavy plant

D. Noise and vibration from the work

**5.26** You are carrying out a noisy work activity and realise that it cannot be finished within the normal working hours of your site. What is the **first** thing you should do?

A. Carry on so that you can finish doing the job as soon as possible

B. Visit the neighbours of the site to tell them what you will be doing

C. Ensure you are wearing appropriate hearing protection before you resume work

D. Stop work and inform site management so they can look at the impact of the activity

**5.27** Why is it **bad** practice to store heavy materials underneath a tree?

A. The tree branches could get damaged

B. Materials are not protected from the tree sap

C. Mould could grow on the stored materials

D. Compaction of the soil could damage the tree roots

**5.28** Which of the following would help to protect the environment?

A. Keeping accurate time sheets

B. Arriving on time for work every day

C. Keeping to the health and safety rules

D. Saving water and energy wherever possible

**5.29** What is the **best** way to minimise dust on site?

A. Covering the whole site

B. Using powered tools only

C. Reducing use of the wheel-wash

D. Dampening using fine water sprays

**5.30** What type of pollution would you associate with hand-held power tools?

A. Smoke

B. Noise

C. Water

D. Light

CONTENTS

# Occupational health

**06**   Dust and fumes (Respiratory hazards)   42
**07**   Noise and vibration   53
**08**   Health and welfare   59
**09**   Manual handling   72

# OCCUPATIONAL HEALTH

## 06    Dust and fumes (Respiratory hazards)

- Harmful dust is often invisible to the naked eye.

- Breathing in harmful dust can cause life shortening illnesses.

- Breathing in construction dust can result in occupational lung diseases, such as asthma and silicosis.

- Respiratory protective equipment (RPE) will only be effective if it fits the wearer's face properly.

- Face-fit testing should be carried out as part of the initial selection of RPE.

- A face-fit test will ensure that your RPE fits and functions properly.

- Wearing your RPE will help to prevent you from breathing in harmful dust and fumes.

- An on-tool extraction system is a method of dust control.

- Fumes will build up very quickly in a confined space.

- Carbon monoxide is a colourless, odourless, poisonous gas.

- Sparks or naked flames can easily ignite flammable vapours.

| 6.01 | You have been asked to do some work that will create dust. What **should** you do? |
|------|-----------------------------------------------------------------|
| A | You should not do the work because dust is highly dangerous |
| B | Start the work. No controls are needed as dust cannot cause serious harm or injury |
| C | Work for short periods at a time. Regular breaks will reduce the amount of dust you breathe in |
| D | Use equipment to eliminate or reduce the dust and wear the correct personal protective equipment (PPE) |

**6.02** If someone is using a petrol cut-off saw (disc cutter) to cut concrete blocks near to pedestrians, what **two** immediate hazards will affect the pedestrians?

You will be asked to 'drag and drop' your answers

(A) Harmful dust

(B) An electric shock

(C) Flying fragments

(D) Contact dermatitis

(E) Vibration white finger

**06**

---

**6.03** Which **two** materials are **most** likely to release silica dust when being cut with a rotating blade?

(A)  Paving slabs

(B)  Concrete blocks

(C)  Timber

(D) Loft insulation

(E) Plastic pipes

**6.05** Where are you likely to be exposed to the highest quantities of dust when drilling, cutting, sanding or grinding?

(A) Inside a small room

(B) Inside a large space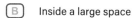

(C) Outside on a still day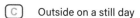

(D) Outside on a windy day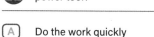

---

**6.04** What is the **main** cause of long-term health issues in the construction industry?

(A) Slipping and tripping

(B) Exposure to loud noise

(C) Being struck by a vehicle

(D) Breathing in hazardous dust and fumes

**6.06** What is the **best** way to limit exposure to dust when using a power tool?

(A) Do the work quickly

(B) Stop dust getting into the air

(C) Stand downwind of any dust

(D) Use the tool during wet weather

**6.07** Which of the following activities does **not** create harmful silica dust?

A  Sawing timber or plywood

B  Breaking up concrete floors and screeds

C  Cutting kerbs, stone, paving slabs, bricks and blocks

D  Chasing out walls and mortar joints or sweeping up rubble

**6.08** After asbestos, which of the following causes the **most** ill health to construction workers?

A  Silica dust

B  Diesel fumes

C  Wood and MDF dust

D  Resin, solvent and paint vapours

**6.09** What is the **main** risk to this worker, wearing **only** these items of personal protective equipment (PPE)?

A  Breathing in harmful dust

B  Back injury, from poor posture

C  Goggles misting up, limiting vision

D  Not being able to hear colleagues

**6.10** What is the **main** risk to this worker, wearing **only** these items of personal protective equipment (PPE)?

A  Breathing in harmful dust

B  Back injury, from poor posture

C  Goggles misting up, limiting vision

D  Not being able to hear colleagues

**6.11**  What is the **most** serious risk to this worker, wearing **only** these items of personal protective equipment (PPE)?

| A | Breathing in harmful dust |
| B | Back injury, from poor posture |
| C | Goggles misting up, limiting vision |
| D | Not being able to hear colleagues |

---

**6.12** Pigeon droppings and nests are found in an area where you are required to work. What **should** you do?

| A | Stop work, do not touch anything, and seek advice |
| B | Carry on with your work carefully, so you don't disturb them |
| C | Try to catch the pigeons so you can move them out of the way |
| D | Wait for the pigeons to fly away before carrying on with your work |

**6.13** How **should** water be used to reduce the level of dust when cutting concrete using a cut-off saw?

| A | Enough to wet the surface of the concrete before cutting |
| B | Constantly the whole time the concrete is being cut |
| C | Constantly until you are half way through the concrete cut, then stop |
| D | Enough to make the first cut, then no more will be required |

**6.14** Before clearing up some rubble **inside** a building, why is it advisable to spray water on it?

| A | So it doesn't make a mess |
| B | To prevent dust clouds |
| C | To kill any insects in it |
| D | To save time |

**6.15** What **should** you do if you find lots of old bird nests and droppings in an area you are working in?

| A | Carry on working and work around them |
| B | Sweep them up and put them in a bin liner immediately |
| C | Stop working and speak to a supervisor to arrange for decontamination work |
| D | Check there are no live birds present, then carry on working |

**06**

**6.16** What **best** describes how workers **should** treat dust?

A. Assume dust is safe if they are working outdoors

B. Assume dust is not safe wherever they are working

C. Assume dust is safe unless told otherwise

D. Assume dust is safe if they don't feel any ill effects

**6.17** Which of the following is **not** an immediate health effect of being exposed to paints and resins which have high levels of solvents?

A. Headaches and sickness

B. Dermatitis or skin problems

C. Muscular and skeletal disorders

D. Drowsiness or poor co-ordination

**6.18** Engine driven equipment is being used in a deep excavation. Which of the following **should** be in place?

A. Gas monitoring alarms

B. Additional excavation covers

C. A vehicle marshal

D. Additional stop blocks

**6.19** Who can enter a confined space?

A. Only competent machine drivers who have the correct licence

B. Anyone who has completed an apprenticeship

C. Only site managers and supervisors of the company

D. Anyone who is trained, competent and authorised

**6.20** Which kind of personal protective equipment (PPE) can protect your lungs from harmful vapours?

A. Goggles

B. Hard hat

C. Respirator

D. Ear defenders

**6.21** Which material or substance is **most** likely to give off hazardous vapour?

A. Dust

B. Rubber

C. Glue

D. Wet concrete

**6.22** What can cause occupational asthma?

A Exposure to rat urine whilst working

B Exposure to loud noise on a regular basis

C Skin contact with any hazardous substance

D Breathing in hazardous dust, fumes or vapours

**6.23** Exposure to which of the following is **unlikely** to result in lung disease?

A Asbestos

B Silica dust

C Strong smells

D Bird droppings

**6.24** Which item of personal protective equipment (PPE) is helping to protect the worker from nasal cancer?

**6.25** Asbestosis is associated with exposure to asbestos. Which part of the body does this disease affect?

A Hands

B Lungs

C Feet

D Brain

**6.26** A construction worker has been exposed to concrete dust for a long period of time. What are they **most** likely to suffer from?

A Headache or migraine

B Blurred vision

C Hearing problems

D Shortness of breath

**6.27** The chances of suffering from lung cancer are **increased** by what?

A Breathing in dust

B Vibration from power tools

C Exposure to sun light

D Exposure to steam

06

**06**

**6.28** When using a power tool to cut or grind materials, why **should** the dust be collected and **not** allowed to get into the air?

A Most dust can be harmful if breathed in

B The tool will go faster if the dust is collected

C To save time and avoid having to clear up the mess

D A machine guard is not needed if the dust is collected

**6.29** What **potential** disease is this worker unprotected from?

A Tetanus

B Nasal cancer

C Dermatitis

D Skin cancer

**6.30** Hydrogen sulphide is a gas given off by rotting organic substances. Which **two** statements are true about hydrogen sulphide?

A It can cause unconsciousness in a few breaths

B It is dangerous because it can disable the sense of smell

C It is a harmless natural gas

D It is dark brown at room temperature

E It can safely be detected by using a naked flame

**6.31** Which **one** of the following is true of repeated exposure to **small** doses of dust?

A It can help to build up immunity

B It is unavoidable and harmless

C Any effects will be immediately apparent

D The effects will build up over time

**6.32** What should you do if you need special respiratory protective equipment (RPE) to handle a chemical but no RPE has been provided?

A Sniff the substance to see if it makes you feel unwell

B Start the work, but take regular breaks to reduce exposure

C Do not start work until you have the correct RPE and training

D Get on with the job, but try to work quickly to reduce exposure

**6.33** If you have been given a dust mask to protect you against hazardous fumes, what **should** you do?

A — Start work without a mask but take regular breaks outside

B — Do the job wearing the mask but work as quickly as you can

C — Do not start work until you have the correct respiratory protective equipment (RPE)

D — Wear a second dust mask on top of the first one, in order to increase the protection

**6.34** The seal between an item of respiratory protective equipment (RPE) and a worker's face is **most** likely to be affected by which **two** of the following?

You will be asked to 'drag and drop' your answers

A — Beard growth

B — Wearing safety goggles

C — Sunlight

D — The wearer's age

E — Dust levels

**6.35** Which **two** factors determine the appropriate type of respiratory protective equipment (RPE) to be used for a job?

A — Whether the RPE is made of rubber or plastic

B — The amount of time since a hazardous spill

C — Whether the worker wants to wear RPE or not

D — The amount of hazardous substances in the air

E — The type of hazardous substance

**6.36** Which **two** of the following are basic filter types used in respiratory protective equipment (RPE)?

A — Moisture filters

B — Smell or aroma filters

C — Sound filters

D — Dust or particle filters

E — Gas or vapour filters

**6.37** Which **one** of the following statements about respiratory protective equipment (RPE) is **true**?

A — Employers must supply it at cost when it is needed

B — Employers must supply it free of charge when it is needed

C — Workers should provide their own

D — Workers should share the cost with the employer

**06**

**6.38** How **should** contaminated respiratory protective equipment (RPE) be considered when being disposed of?

A. As recyclable materials

B. As normal waste products

C. As compostable wastes

D. As hazardous waste

**6.39** A particle filter is suitable for use in which **one** of the following situations?

A. Presence of gases

B. An oxygen-deficient atmosphere

C. When dust and fibres are in the air

D. Presence of vapours

**6.40** Why is it important to be clean shaven when using a half-mask respirator?

A. Facial hair can block the filter more quickly

B. You may suffer an allergic reaction to the mask

C. Facial hair can affect the seal around your face

D. You will be able to use the same mask for longer

**6.41** Which of the following do you **not** need to do to ensure that someone's mask works?

A. Check the mask is being worn correctly

B. Check the mask is the correct type needed

C. Check the mask under water to make sure the seals are tight

D. Check the user has passed a face-fit test while wearing the mask

**6.42** The seal of your respiratory protective equipment (RPE) can be broken by which **two** things?

You will be asked to 'drag and drop' your answers

A. Facial hair

B. Facial scarring

C. Make-up

D. A hearing aid

E. Earrings

**6.43** What is the **most** important consideration when wearing respiratory protective equipment (RPE)?

A The weight is correct

B It has a good seal

C Being able to smell

D It is the correct colour

**6.44** If your respiratory protective equipment (RPE) is a bad fit, which **one** of the following is **most** likely to happen?

A It will not protect you

B It will break easily

C It will filter more air

D It will get damaged

**6.45** How often is it good practice to carry out repeat face-fit tests for respiratory protective equipment (RPE)?

A On a regular basis

B On an ad-hoc basis

C When starting a new shift pattern

D When starting work on a different site

**6.46** Respiratory protective equipment (RPE) fit tests **should** be carried out by whom?

A The worker who will carry out the work

B A supervisor, in compliance with the law

C The manager overseeing the work

D A competent person in compliance with the law

**6.47** Planned work requires the use of a power tool to cut or grind materials. Select the **two** **best** ways to control the dust.

You will be asked to 'drag and drop' your answers

A Wet cutting

B Wear a dust mask

C Work slowly and carefully

D Keep the area clean and tidy

E Fit a dust extractor or collector to the machine

06

**6.48** What **must** you do when using water to keep dust down when cutting?

A) Ensure that there is as much water as possible

B) Make sure that the water flow is correctly adjusted

C) Pour water onto the surface before you start cutting

D) Get someone to stand next to you and pour water from a bottle

**6.49** When drilling, cutting, sanding or grinding, what is the **best** way to protect your long-term health from harmful dust?

A) Use dust extraction, or wet cut and wear a dust mask

B) Wear FFP3-rated respiratory protective equipment (RPE)

C) Wear any disposable respiratory protective equipment (RPE)

D) Use dust extraction, or wet cut and wear FFP3-rated respiratory protective equipment (RPE)

**6.50** What should you do if you run out of the water you are using to control dust?

A) Stop and refill the water

B) Put on additional respiratory protection

C) Ask everyone to clear the area and then carry on

D) Carry on but get someone to sweep up afterwards

**6.51** Which of the following **two** options are likely to cause the **most** dust exposure?

A) Using power tools without extraction

B) Using hand tools outside

C) Working with wet or damp materials

D) Working with dry materials

E) Using power tools with extraction

**6.52** When working with materials creating dust, what **should** be monitored?

A) The level of exposure to the dust

B) The colour of dust created

C) The smell the dust creates

D) The direction in which the dust travels

**6.53** Using water suppression to reduce dust will be **most** effective for which **one** of the following?

A) Steel grinding

B) Cutting plywood sheets

C) Disc cutting steel

D) Pneumatic chiselling of concrete

## 07    Noise and vibration

- If you have to raise your voice to be understood as a result of noise on site, stop work and raise the problem with your supervisor.

- Always wear the correct protection in a hearing protection zone.

*Hearing protection must be worn*

07

- Exposure to vibration is a serious issue as it can result in disabling health conditions that cannot be cured.

- Hand-arm vibration syndrome (HAVS) includes a range of conditions that can lead to permanent damage in the hands and forearms.

- Regular use of hand-held tools and equipment that vibrates is the main cause of hand-arm vibration syndrome.

- Your employer should explain safe methods of use, and give you advice on exposure times for hand-held vibrating tools.

---

**7.01**    How can excessive noise levels affect your hearing? Give **two** answers.

You will be asked to 'drag and drop' your answers

| A | Hearing improvement |
|---|---|
| B | Ear infections |
| C | Permanent hearing loss |
| D | Temporary hearing loss |
| E | Dizziness and nausea |

**7.02** What are the signs and symptoms of noise-related hearing damage?

A | Ear infections and regular headaches

B | Nausea and a skin rash around your ears

C | There are no signs or symptoms associated with hearing damage

D | Difficulty following a conversation, especially against background noise

**7.03** If you hear a ringing sound in your ears after working with noisy equipment, what does this mean?

A | The noise level was high but acceptable

B | You have also been subjected to vibration

C | Your hearing has been temporarily damaged

D | Your hearing protection was working properly

**7.04** Can the damage by exposure to noise over a long period of time be reversed?

A | Yes, if you change jobs

B | Yes, if you have an operation

C | No, the damage is permanent

D | No, unless medication is used

**7.05** What should an employee do if they think noise at work may have damaged their hearing?

A | Take time off work, as they are unwell

B | Nothing, as the damage has already been done

C | Ask their employer or doctor to arrange a hearing test

D | Plug their ears with cotton wool to stop any more damage

**7.06** What is the **main** risk to this worker, wearing **only** these items of personal protective equipment (PPE)?

A | Dermatitis to skin

B | Damage to hearing

C | Eye injuries

D | Breathing in harmful dust

**7.07** Your doctor tells you that you have hand-arm vibration syndrome (HAVS), possibly caused through work. What **should** you do?

A  Tell no one, as it's not contagious

B  Only inform your friends at work

C  Inform your supervisor or employer

D  Tell no one, as HAVS is not reportable

**7.10** What health problem can be caused by using hand-held vibrating tools?

A  Blisters on your fingers and hands

B  Skin cancer on your hands and arms

C  Damage to the blood vessels in your fingers and hands

D  An itchy skin irritation, like dermatitis, affecting your hands

07

**7.08** What are **three** early signs of hand-arm vibration syndrome (HAVS)?

You will be asked to 'drag and drop' your answers

A  Rash on the fingers

B  Fingertips turn white

C  Blisters on the fingers

D  Temporary loss of feeling in the fingers

E  Tingling or a pins and needles sensation in the fingers

**7.09** Which one of these tools is **most** likely to cause hand-arm vibration syndrome (HAVS)?

A  Handsaw

B  Hammer drill

C  Hammer and chisel

D  Battery-powered screwdriver

**7.11** You are **less** likely to suffer from hand-arm vibration syndrome (HAVS) if you feel which one of the following?

A  Cold but dry

B  Cold and wet

C  Warm and dry

D  Wet but warm

# OCCUPATIONAL HEALTH

**7.12**  You have been using a vibrating tool and the ends of your fingers are starting to tingle. What does this mean?

**A** You can carry on using the tool but you must hold it more tightly

**B** You can carry on using the tool but you must loosen your grip

**C** You need to report your symptoms before they cause a problem

**D** You must not use this tool, or any other vibrating tool, ever again

**7.15** What does wearing hearing protection do?

**A** Helps you to hear better

**B** Repairs your hearing if it is damaged

**C** Stops you hearing all noise in the workplace

**D** Reduces damaging noise to an acceptable level

---

**7.13**  What are **two** recommended ways to protect your hearing?

You will be asked to 'drag and drop' your answers

**A** Earplugs in your ears

**B** Ear defenders over your ears

**C** Soft cloth pads over your ears

**D** Rolled tissue paper in your ears

**E** Cotton wool pads over your ears

---

**7.14**  Noise levels may be a problem if you have to raise your voice to be heard by someone standing how far away?

**7.16** If you need to wear disposable foam earplugs, how should you insert them so they protect your hearing from damage?

**A** Do not roll or fold them, and insert them half way into your ear canal

**B** Fold them in half, pull on your earlobe and wedge them half way into your ear

**C** Soak them in water, squeeze them out and then insert them into your ear canal

**D** Roll them up and insert them, while pulling the top of your ear up to open up the ear canal

**7.17** What **should** you do if you need to wear ear defenders but an ear pad is missing from one of the shells?

A. Put them on and work with them as they are

B. Do not work in noisy areas until they are replaced

C. Take an ear pad from another set of ear defenders

D. Leave them off and work without any hearing protection

**7.20** How can you help reduce the risk of hand-arm vibration when using a vibrating tool?

A. Hold the tool more tightly

B. Use more force on the tool

C. Hold the tool at arm's length

D. Do not grip the tool too tightly

**07**

**7.18** Using a grinder whilst wearing this personal protective equipment (PPE), **could** result in which of the following?

A. Lung disease

B. Weil's disease

C. Hearing damage

D. Eye injuries

**7.19** What **should** you do if someone near you is using noisy equipment and you have **no** hearing protection?

A. Speak to the other person's supervisor to stop them making the noise

B. Ask them to stop what they are doing, as it is disrupting other workers on site

C. Carry on with your work, as you are not the person using the noisy equipment

D. Leave the area until you have the correct personal protective equipment (PPE)

**7.21** How can the effects of hand-arm vibration be reduced if you are using vibrating tools?

A. Complete the job in one long burst

B. Only use one hand at a time on the tool

C. Do not smoke, as it affects blood circulation

D. Hold the tool as tightly as you can and work quickly

**7.22**    What risk does this worker face by using this tool on a **regular** basis?

A    Dermatitis from repetitive use

B    Weil's disease caused by the dust

C    Hand-arm vibration from prolonged use

D    Back injury caused by using heavy equipment

**7.23**    What do the initials **HAVS** stand for?

A    Hand-arm vibration syndrome

B    Hand and ventilation system

C    Heavy arm vibration system

D    Heat and ventilation syndrome

**7.24**    Which **two** potential health issues are more likely when using a hammer drill for long periods of time?

A    Carpal tunnel syndrome

B    Vibration white finger

C    Hepatitis

D    Head injuries

E    Speech impairment

## 08    Health and welfare

- If you are under the influence of alcohol, drugs or prescribed medication, it can make you feel drowsy, slow your reaction times and affect your judgement. This will increase the risk of an accident at work.

- Long working hours or poorly designed shift work schedules can result in fatigue. If you are suffering from fatigue, you are more likely to have an accident at work.

- Fatigue can result in an increase in errors in the workplace.

- Stress is defined as the adverse reaction people have to excessive pressures or demands placed on them.

- Stress can affect anyone. A lack of concentration, anger and sleep problems are all warning signs.

- Stress at work can have a negative effect on your mental wellbeing.

- If you are concerned about a colleague's mental health, speak to them about it.

- Talking and listening to people, without judgement, can help to overcome negative attitudes to mental health in the workplace.

- If you feel that you are having mental health issues, you should ask for help as early as possible.

- Talking about mental health issues is a good way of helping to manage them.

*Talking about how you feel is not a sign of weakness*

- Exposure to ultraviolet radiation from the sun is one of the main causes of skin cancer in the construction industry.

- You are less likely to have an accident if your work area is clean and tidy.

- Adequate lighting and good housekeeping will reduce the risk of slips and trips.

- If the welfare facilities on site are not adequate or are dirty, report the issue to your supervisor.

- It is good practice to have access to a telephone and a means of raising the alarm if you are working alone.

08

**08**

**8.01** A worker is taking medication which **could** affect their health and safety in the workplace, and that of others. What **should** they do?

A) Tell their manager

B) Work harder on site

C) Go and see their doctor

D) Take regular breaks

**8.02** When an employee returns to work after an absence due to illness, what **should** they speak about with their employer?

A) The details of the illness to make sure no one else can catch it

B) The effects of medication they are taking which could affect safety at work

C) The number of times they have visited their doctor before returning to work

D) Any treatment costs they have to pay to help them get back to work

**8.03** What action **should** shift workers take at work if they are taking time-dependent medication, such as insulin?

A) Ask colleagues to help them remember when to take their medication

B) Consult their doctor and inform their manager to help plan how to accommodate this

C) Only work shifts after they have taken their required medication

D) Not tell their colleagues each time they need to take their medication during their shift

**8.04** How should absence records containing specific medical information relating to an employee be treated?

A) Confidentially, in accordance with data protection laws

B) As public information, in accordance with freedom of information

C) The records can be shared with the workforce as long as they do not say anything

D) Provided to the rest of the workforce to prevent others taking time off

**8.05** If your doctor has given you some medication, which of these questions is the **most** important to ask?

A) Will I fail a drugs test if my employer asks for one?

B) Will it cause me to oversleep and be late for work?

C) Will it make me unsafe to work or operate machinery?

D) Will it make me work more slowly and earn less money?

**8.06** If you suspect someone at work has been drinking alcohol, what **should** you do?

A) Ask them to stay away for an hour and then go back to work

B) Get them to drink plenty of strong coffee before they go back to work

C) Report the situation to your supervisor, as they may be unsafe to work

D) Get them to eat and drink something, wait 30 minutes and then go back to work

**8.07** What are the minimum facilities that **must** be provided on site for washing your hands?

A   A cold water standpipe and paper towels

B   A water container, bowl and paper towels

C   There is no need to provide washing facilities

D   Hot and cold water, soap and a way to dry your hands

**8.08** What are **two** ways of reducing the risk of transferring hazardous substances from your hands to your mouth?

A   Washing your hands before eating

B   Using barrier cream for working activities

C   Washing protective gloves before each use

D   Wearing protective gloves while you are working

E   Putting barrier cream on your hands before eating

**8.09** What **should** you use to clean very dirty hands?

A   Paraffin

B   Thinners

C   White spirit

D   Soap and water

**8.10** What should you do if there is nowhere on site to wash your hands?

A   Wait until you get home, then wash them

B   Go to the local public toilets and use their washbasin

C   Nothing, as the site does not have to provide washing facilities

D   Speak to your supervisor or the site manager about the problem

**8.11** Why should you **not** use white spirit or other solvents to clean your hands?

A   They could block the pores of the skin

B   They will remove several layers of skin

C   They could strip the protective oils from the skin

D   They could carry harmful bacteria that attack the skin

**8.12** What is the **main** issue with using barrier cream to protect your skin?

A   It is difficult to wash off

B   It costs too much to use every day

C   It can be broken down by some substances

D   It can irritate your skin and give you dermatitis

08

**8.13** When **should** you apply barrier cream to your skin?

A. Before you start work

B. When you finish work

C. As part of first-aid treatment

D. When you can't find your gloves

**8.14** What can cause occupational dermatitis?

A. Using tools that vibrate

B. Working in the sun without sun cream

C. Contact with another person who has dermatitis

D. Contact with some strong chemicals or substances

**8.15** What condition can be caused by direct sunlight on bare skin?

A. Acne

B. Rickets

C. Dermatitis

D. Skin cancer

**8.16** To help protect outdoor workers from the risk of skin cancer from sun exposure, what **should** be worn?

A. Low factor tanning oil and short-sleeved shirts

B. High factor sunscreen and long-sleeved clothing

C. Extra moisturiser on the face and short-sleeved shirts

D. A small amount of deodorant and vest tops

**8.17** Prolonged exposure to sunlight **could** cause what?

A. Hair loss

B. Burns

C. Abrasions

D. Dental issues

**8.18** When referring to protection, what is a high UV rate cream designed to protect you from?

A. Dermatitis

B. Legionella

C. Sun burn

D. Abrasions

08

**8.19** What is the **most** likely source of hepatitis in this image?

**8.20** How does tetanus (an infection that you can catch from contaminated land or water) **normally** enter your body?

(A) Through the pores in your skin

(B) Through an open cut in your skin

(C) Through your nose when you breathe

(D) Through your mouth when you eat or drink

**8.22** Reducing the risk of cuts and abrasions would require protection for what part of the body?

(A) Bones

(B) Blood

(C) Hair

(D) Skin

**08**

**8.23** Which disease is most likely to be caught through cuts, grazes or puncture wounds?

(A) Tetanus

(B) Dermatitis

(C) Legionella

(D) Cancer

**8.21** What is the **main** risk to this worker, wearing **only** these items of personal protective equipment (PPE)?

(A) Damage to hearing

(B) Eye injuries

(C) Breathing in harmful dust

(D) Cuts and abrasions to skin

# OCCUPATIONAL HEALTH

**8.24** Select the **two** images in which the worker is correctly protecting themselves from possible cuts or abrasions.

You will be asked to 'drag and drop' your answers

- (A) Carrying with no gloves
- (B) Washing with no gloves
- (C) Scraping with gloves
- (D) Scrubbing with no gloves
- (E) Painting with gloves

---

**8.25** Which animal is the **most** likely carrier of Weil's disease (leptospirosis) on construction sites?

- (A) Rabbits
- (B) Rats
- (C) Squirrels
- (D) Cats

**8.26** If your doctor says that you contracted Weil's disease (leptospirosis) on site, why do you need to tell your employer?

- (A) Your employer has to warn your colleagues not to go near you
- (B) It must be reported to the Health and Safety Executive (HSE)
- (C) The site on which you contracted it will have to be closed down
- (D) Your employer will need to call pest control to remove rats on site

**8.27** In what situation are you **most** likely to catch Weil's disease (leptospirosis)?

- (A) If you drink water from a standpipe
- (B) If you work fixing showers or baths
- (C) If you work near air-conditioning units
- (D) If you work near wet ground, waterways or sewers

**8.28** What other illness can be easily confused with the early signs of Weil's disease (leptospirosis)?

- (A) Diabetes
- (B) Hay fever
- (C) Dermatitis
- (D) Influenza (flu)

**8.29** Fatigue could affect work rates of a worker. What does this mean?

A) They will be able to work faster as they will have lots of energy

B) They will work consistently as there are no issues

C) They will work at the same rate but will need monitoring

D) They will work more slowly, as they will feel tired

**8.30** What type of work is **most** likely to result in fatigue?

A) New and challenging

B) Repetitive and monotonous

C) Exciting and enjoyable

D) Quiet and interesting

**8.31** Being satisfied with your job can lead to what?

A) Feeling less stressed at work

B) Feeling unpopular at work

C) Feeling more stressed at work

D) Feeling less happy at work

**8.32** Which **one** of the following is **most** likely to cause stress at work?

A) Job satisfaction but fear of redundancy

B) A lack of job security and fear of redundancy

C) Job security and a permanent contract

D) A lack of job security but a permanent contract

**8.33** How can physical stress of a job be reduced?

A) Repetitive actions when working

B) Job rotation and task variation

C) Making equipment challenging to use

D) An increase in pay for the same job

**8.34** If a worker is feeling symptoms or showing signs of stress at work, what **should** they do?

A) Speak to someone they trust, like a friend or someone independent

B) Arrive and start work at a later time

C) Tell the rest of the team about their problems

D) Eat or drink more during the day

**8.35** Which statement about mental health is **true**?

A  It always has an obvious cause

B  It can have no obvious cause

C  It is always caused by stress

D  It is never caused  by stress

**8.36** Who has a duty to protect an individual from stress at work?

A  The union

B  The government

C  The local authority

D  The employer

**8.37** What is the organisation 'The Samaritans'?

A  A charity that provides emotional support for people who are struggling to cope

B  A political party which supports looking after vulnerable people

C  A  trade union organisation offering financial and legal support

D  A charity offering construction training to young vulnerable people

**8.38** Loss of appetite, fatigue and tearfulness are common symptoms of what?

A  Mental health issues or stress

B  Hand arm vibration syndrome

C  Repetitive strain injury and back pain

D  Skin problems such as dermatitis

**8.39** Which **one** of the following statements about mental health is **true**?

A  Mental and physical health are directly linked

B  Mental health is all about our intelligence

C  Mental health has to do with our general knowledge

D  Mental health is all about how we think and remember things

**8.40** Which **one** of the following statements is **true**?

A  People experiencing mental health problems tend to be violent or dangerous

B  Mental health problems are common and can happen to anyone

C  Learning difficulties and mental health problems are the same

D  Mental health problems are rare among construction workers

**8.41** Which **one** of the following is good advice for helping to cope with stress?

A Get enough rest

B Drink more alcohol

C Work longer hours

D Keep it to yourself

**8.42** A worker is suffering stress caused by their line manager. What **should** they do?

A Get another job that will be less demanding

B Try to work faster to keep the manager happy

C Find and follow the company procedures to address it

D Complain to their manager

**8.43** Which **one** of the following is true of the symptoms of stress?

A They are the same for everybody

B They can be different for each individual

C They always develop very quickly

D They always take a while to develop

**8.44** MIND is a charity that does what?

A Provides advice and support to empower anyone experiencing a mental health problem

B Provides housing for retired construction workers

C Represents people who are very intelligent and want to improve their IQ

D Controls and monitors health and safety in the work place

**8.45** What sort of rest area should your employer provide on site?

A A canteen serving food, drinks and cold sandwiches

B A covered area, chairs, and a way to boil water and heat food

C A covered area with some comfortable chairs and running water

D Employers don't have to provide rest areas, as long as rest breaks are provided

**8.46** Which **one** of the following must be provided on site?

A Snacks

B Drinking water

C Free transport home

D Breakfast

08

**8.47** Rest areas on site **should** be equipped with which of the following?

A Settees and chairs

B Televisions and tables

C Seating and radio

D Seating and tables

**8.48** How can everyone on site help keep rats away?

A Put rat traps and poison around the site

B Ask the Local Authority to put down rat poison

C Throw food scraps over the fence or hoarding

D Put all food and drink rubbish into bins provided

**8.49** Which of the following is **true** of clearing waste to maintain a tidy site?

A Clearing waste should be carried out at the end of a shift

B Construction workers are not responsible for clearing waste

C Construction sites are dirty anyway and do not need clearing

D Clearing waste should be a continuous process

**8.50** When absorption granules from a spill kit have been used on oil, what action should be taken?

A Clear them up and place them in a sealed waste bag ready for specialist disposal

B Leave them on the oil for a few days before clearing into the general waste

C Clear them up straight away and put them into a general waste skip

D Use water to help clean up excess oil before specialist disposal

**8.51** When **should** an oil spill be cleaned up?

A At the end of the shift

B When it has dried

C Never – it will be absorbed into the ground

D Immediately – it could cause someone to slip

**8.52** What is the **most** important reason for keeping your work area clean and tidy?

A To help prevent slips, trips and falls

B To recycle waste and help the environment

C So that waste skips can be emptied more often

D It saves time cleaning up at the end of the week

08

**8.53** A worker creates offcuts on site. Who is responsible for clearing them away?

A  The worker

B  The supervisor

C  The site manager

D  The foreman

**8.54** The ground has become muddy on site. What **could** be done to prevent the ground becoming slippery?

A  Treat the surface with salt

B  Treat the surface with gravel

C  Improve lighting

D  Improve signage

**8.55** A work task results in cables from power tools running across a walkway. What action **should** be taken?

A  While working, look out for anyone approaching to warn them

B  Think about cancelling the job because it is too dangerous

C  Consider using cordless tools, or running the cables at high level

D  Put up signs that the fire escape is out of order temporarily

**8.56** Fatigue may be a result of what?

A  Good work/life balance

B  Good sleeping patterns

C  A healthy diet

D  Working long hours

**8.57** What would be a good way of reducing fatigue in the workforce?

A  Regular start and finish times

B  Early start times and a late finish

C  Rotating shift patterns

D  Random start and finish times

**8.58** What can help to reduce fatigue?

A  Going to the gym less

B  Taking regular breaks at work

C  Drinking alcohol after work

D  Eating larger meals during break times

**08**

**8.59** If a worker is feeling stressed, when is the **best** time for them to address the issue?

A When they have finished work and they are away from the workplace

B As soon as they realise they have symptoms of stress

C Only after the stress level gets so bad it causes an accident

D In about six months, if the issue is still causing them stress

**8.60** What is the best way for a worker to avoid becoming stressed because of an overload of work?

A Speak openly and regularly with their manager or employer about workloads

B Put up with the extra work but make sure overtime is paid

C Only do what is manageable because someone else will pick up the extra

D Make sure they take medication before going to work

**8.61** What is **one** sign that employees are feeling stressed at work?

A Increased productivity on site

B Fewer accidents on site

C Long-term staff retention rates

D High staff turnover rates

**8.62** Which of the following can be an indicator of stress?

A Inability to deal with usual workload

B Increased productivity

C Feeling valued at work

D Feelings of confidence at work

**8.63** Which of the following are common mental health issues?

A Paralysis and halitosis

B Hand arm vibration syndrome

C Depression and anxiety

D Dermatitis, skin irritation

**8.64** If a worker confides in a colleague that they have suffered from a mental health issue, what **should** the colleague do?

A Let other colleagues know, so they can avoid working with them

B Treat them as they would any other work colleague

C Inform the site supervisor and first aider

D Do their work for them because they might not be able to cope

**8.65** When might people suffering with mental health issues need help at work?

A. When their site supervisor or manager says so

B. They will regularly need help to avoid them having a panic attack

C. Up to once a week to take some of the strain off them

D. As and when a situation arises in which help is needed

**8.66** Lone workers are most a risk from what?

A. Violence

B. Paranoia

C. Sleeplessness

D. Humiliation

**8.67** Who should drive company vehicles?

A. Anyone with a learner driver permit

B. Any employee who is competent and authorised

C. Any construction site manager or supervisor

D. Any junior apprentice workers

**8.68** A worker with a full UK driving licence has been asked to move a machine they have never been trained on. What should the worker do?

A. They can move the machine as they have a full UK driver's licence

B. Move the machine as long as there is no one else near it

C. Explain that they are not trained and competent to move it

D. Move the machine as long as there is a vehicle marshaller

08

## 09    Manual handling

- Using a wheelbarrow or other lifting aids to move heavy loads is classed as manual handling but they help to reduce the risk of personal injury.

- Workers should be trained in safe lifting techniques before manual handling or lifting.

- Adopting safe manual handling techniques will help to protect your back and reduce the risk of injury in the workplace.

- The manual handling acronym T.I.L.E. stands for Task, Individual, Load, Environment.

*Good practice method for kinetic lifting*

**9.01** Your new job involves some manual handling but an old injury means that you have a weak back. What **should** you do?

- A  Tell your supervisor you can lift anything on site

- B  Tell your supervisor that lifting might be a problem

- C  Try some lifting then tell your supervisor about your back

- D  Tell your supervisor about your back if it gets injured again

**9.02** You have to move a load that might be too heavy for you. You cannot divide it into smaller parts and there is no-one to help you. What should you do?

- A  Try to lift it using the correct lifting methods

- B  Lift and move the load quickly to avoid injury

- C  Do not move the load until you have a safe way of doing it

- D  Get a forklift truck, even though you have not been trained to use it

**9.03** You need to lift a load that is not heavy, but it is so big that you cannot see in front of you. What **should** you do?

A Get someone to walk next to you and give directions

B Ask someone to help carry the load so that you can both see ahead

C Get someone to walk in front of you and tell others to get out of the way

D Move the load on your own. It is so large that anyone in your way is sure to see it

**9.04** What **should** you do if you need to carry a load down a steep slope?

A Carry the load on your shoulder

B Assess whether you can still carry the load safely

C Walk backwards down the slope to help you balance

D Put the load down and let gravity move it down the slope

**9.05** What are **two** risks of carrying a load in cold, damp conditions?

A The load will be easier to carry

B The route you take could be slippery

C You will need to work more quickly to warm up

D The load will feel lighter due to the cold conditions

E Your ability to carry the load safely will be reduced

**9.06** What **should** you do if you have been told how to lift a heavy load but you think there is a better way to do it?

A Discuss your idea with your supervisor before lifting

B Ignore what you have been told and do it your way

C Forget your idea and do it the way you have been told

D Ask your workmates to decide which way you should do it

**9.07** What does it mean if you have to twist or turn your body when you lift and place a load?

A You must wear a back brace in this situation

B You will be able to lift the same weight as usual

C The weight you can lift safely will be less than usual

D The weight you can lift safely will be more than usual

**9.08** If you need to reach above your head to place a load or lower a load to the floor, which of these is **not** true?

A The load will be more difficult to control

B You can safely handle more weight than usual

C It will be more difficult to keep your back straight

D You will put extra stress on your arms and your back

09

**9.09** If you have to move a load while you are sitting down, how much can you lift safely?

A. The usual amount

B. Twice the usual amount

C. Less than the usual amount

D. Three times the usual amount

**9.10** Which part of the body is **most** likely to be injured when lifting heavy loads?

**9.11** Which part of the body is **most** likely to be injured when lifting heavy loads?

**9.12** What is the outcome of wearing a back support belt when lifting?

A. You can safely lift more than usual

B. You could face the same risk of injury

C. You can lift any load without being injured

D. You will crush your backbone and damage it

**9.13** Which **three** of the following factors must you think about to lift a load safely?

You will be asked to 'drag and drop' your answers

A. Its weight

B. Its size and shape

C. What the value of it is

D. How to grip or hold it firmly

E. Whether the contents are insured

**9.14** What **two** things are important for the use of manual handling lifting aids?

You will be asked to 'drag and drop' your answers

A The user must hold a CSCS card

B The lifting aid can only be used outside

C The lifting aid must be designed for the task

D Lifting aids must not be more than six months old

E Users must be trained in the correct use of the lifting aid

**09**

**9.15** You need to move a load that might be too heavy for you. What **three** methods could you use?

You will be asked to 'drag and drop' your answers

A Ask someone to help you

B Drag the load to avoid lifting it

C Use an aid, such as a trolley or wheelbarrow

D Divide the load into smaller loads if possible

E Test the load's weight by picking it up for a short time

**9.16** What **must** all workers do under the regulations for manual handling?

A Follow their employer's safe systems of work

B Wear back-support belts when lifting things at work

C Make a list of all the heavy things they have to carry

D Lift any size of load once the risk assessment has been done

**9.17** If you need to move a load that is heavier on one side than the other, how should you pick it up?

A With the heavy side towards you

B With the heavy side away from you

C With the heavy side on your weak arm

D With the heavy side on your strong arm

# OCCUPATIONAL HEALTH

**9.18** Which of the following is the **best** method to help minimise the risk of injury when moving loads on site?

A. Ask a trained person to carry the load

B. Use lifting aids wherever possible on site

C. Make the area flatter before performing the task

D. Remove all awkward shaped loads from the site

**9.19** A wheel comes off a trolley you are using to move a heavy load a long distance. What **should** you do?

A. Carry the load for the rest of the journey

B. Drag the trolley on your own for the rest of the journey

C. Find another way to move the load and complete the journey

D. Ask someone to help you pull the trolley for the rest of the journey

**9.20** Who should be involved in planning the safe system of work for your manual handling?

A. You and your colleagues

B. Your supervisor or employer

C. You and your supervisor or employer

D. The Health and Safety Executive (HSE)

**9.21** If you are required to lift a heavy load, what **must** your employer do?

A. Watch you while you lift the load

B. Complete a risk assessment of the task

C. Nothing. Lifting loads is a part of your job

D. Make sure that a supervisor is there to advise while you lift

**9.22** Which one of the following **could** cause back and musculoskeletal problems for a worker?

A. Positioning materials away from the work area

B. Good planning to reduce lifting heavy loads

C. Reducing the maximum lifting weight

D. Using machines for lifting operations whenever possible

# Safety

| | | |
|---|---|---|
| **10** | Safety signs | 78 |
| **11** | Fire prevention and control | 81 |
| **12** | Electrical safety, tools and equipment | 86 |

## 10 Safety signs

● Prohibition signs – must not do – red and white

No access for pedestrians | No admittance – Authorised personnel only | No mobile phones | No naked flames | No smoking

Not drinkable | Do not touch

● Mandatory signs – must do – blue and white

Safety gloves must be worn | Safety boots or shoes must be worn | Safety harness must be worn | Safety helmet must be worn | Protective eyewear must be worn

Hearing protection must be worn | General mandatory sign

● Warning signs – yellow and black

Slippery surface | Toxic material | Trip hazard | Combustible or flammable material

Corrosive material | High voltage | Industrial vehicles operating | Radioactive material

*Explosive material*  *General warning sign*  *Overhead load*

● Safe condition signs – green and white

*Emergency assembly point*  *Emergency escape route*  *Emergency eye wash*  *Emergency shower*  *First aid*

● Fire-fighting signs

*Fire emergency telephone*  *Fire extinguisher*  *Fire fighting equipment*  *Fire hose reel*  *Fire ladder*

*Fire alarm call point*

● Globally harmonised pictograms are used to help identify hazardous substances

*Hazardous to the environment and aquatic life*  *Oxidising gases, liquids and solids*  *Damage to organs*  *Heating may cause an explosion*  *Corrosive*

*Flammable gases*  *Harmful – skin, eye or respiratory irritation*  *Contains gases under pressure*  *Toxic*

10

**10.01** Which image shows the worker correctly following these site safety signs?

A

C

B

D

---

**10.02** Which three of the following should be labelled with this sign?

You will be asked to 'drag and drop' your answers

A

B

C

D

E

| Raw asbestos |
| Asbestos waste |
| Recyclable waste |
| Plasterboard waste |
| Any product containing asbestos |

---

**10.03**

How should a container, or any residue, be disposed of if it has this sign on the label or packaging?

A   Put it in any type of skip or bin

B   Leave it somewhere for other people to deal with

C   If it is a liquid and less than one litre you can pour it down a drain

D   Follow specific instructions on the label and in the work instructions

## 11  Fire prevention and control

- Emergency procedures should be in place before any work begins, explained in the site induction and not changed without notice.

- Everyone on site should be aware of the emergency procedures, as these help to control dangerous situations.

- Emergency escape routes should be kept clear and unobstructed at all times.

- A fire assembly point is where people must go when the fire alarm sounds.

- Hot-work permits authorise tasks to be carried out safely, under strictly controlled conditions.

- Hot-work permits allow workers to carry out work that could start a fire, such as cutting steel with an angle grinder or soldering pipework in a central heating system.

- Liquefied petroleum gas (LPG) is colourless, has a distinctive smell and is highly flammable.

- A leaking LPG cylinder can catch fire at some distance from the original leak and flash back to the source.

- If a cylinder of LPG is leaking, turn the supply off immediately, if it is safe to do so.

- Fuel should always be dispensed using the correct nozzle and stored in the designated fuel store when not being used.

- Refuelling should only be carried out by authorised people, when the equipment is turned off and cooled down for safety.

11

**11.01**   What are two common fire risks on construction sites?

You will be asked to 'drag and drop' your answers

| A | Timber racks |
|---|---|
| B | 230 volt power tools |
| C | Uncontrolled hot works |
| D | 110 volt extension reels |
| E | Poor housekeeping and build up of waste |

# SAFETY

**11.02** In addition to heat, what are the other two factors that must be present to start a fire?

You will be asked to 'drag and drop' your answers

- A  Nitrogen
- B  Carbon dioxide
- C  Argon
- D  Oxygen
- E  Fuel

Heat

**11.03** What must be checked before working in a corridor that is a fire escape route?

- A  If the tools being used are spark-proof
- B  If the doors into the corridor are locked
- C  That any tools and equipment do not block the route
- D  That fire escape signs are removed before the work starts

**11.04** If you discover a fire, what is the first thing that you should do?

- A  Raise the alarm
- B  Put your tools away
- C  Try to put out the fire
- D  Finish what you are doing

**11.05** A large fire has been reported. You have not been trained to use fire extinguishers. What should you do?

- A  Leave work for the day
- B  Go straight to the assembly point
- C  Report to the site office and then go home
- D  Put all of your tools away and then go to the assembly point

**11.06** Which emergency procedures should be explained in the site induction?
Give three answers

You will be asked to 'drag and drop' your answers

A  How to raise the alarm in case of an emergency

B  Where to go if the fire alarm is activated

C  Where to go to leave valuables in an emergency

D  What to do if someone is injured on site

E  How to avoid leaving site in case it's a false alarm

---

**11.07** If a fire occurs, how should you interact with the designated fire warden?

A  Follow the instructions given by the fire warden

B  Ignore the fire warden and follow your colleagues

C  Follow the site manager as they will know their way around the site

D  Ignore the fire warden and the site manager

**11.08** What additional rules might a person working alone be asked to follow?

A  Carry two sets of personal protective equipment (PPE)

B  Complete an additional time sheet

C  Never speak to the general public

D  Make regular radio or mobile phone contact

11

---

**11.09** If your job needs a hot-work permit, what two things would you expect to have to do?

You will be asked to 'drag and drop' your answers

A  Write a site evacuation plan

B  Know how to refill fire extinguishers

C  Check for signs of fire when you stop work

D  Have a fire extinguisher close to the work

E  Know where all the fire extinguishers are kept on site

**11.10** You are about to start a job. How will you know if you need a permit to work?

A The Health and Safety Executive (HSE) will give them out

B Permits to work are only required by managers on large sites

C Information will be given during the site induction before any work starts

D Other workers on site will pass them on after they have finished with them

**11.11** What does it mean if there is frost around the valve on a liquefied petroleum gas (LPG) cylinder?

A The cylinder is full

B The valve is leaking

C The cylinder is nearly empty

D You must lay the cylinder on its side

**11.12** Which two extinguishers should not be used on electrical fires?

A Dry powder (blue colour band)

B Foam (cream colour band)

C Water (red colour band)

D Carbon dioxide (black colour band)

**11.13** Match the fire extinguisher with the described contents

You will be asked to 'drag and drop' your answers

Water – for use on wood, paper, textile and solid material fires

Powder – for use on liquid and electrical fires

Foam – for use on liquid fires

Carbon Dioxide ($CO_2$) – for use on liquid and electrical fires

**11.14** What is the primary purpose of fire extinguishers?

A. To tackle small fires to prevent them becoming larger

B. To be available and ready for when the fire services arrive

C. To make sure the premises pass a safety inspection

D. To add decoration to the walls of the construction site

**11.15** A worker needs to use a flammable liquid. How much should be taken from the store?

A. Enough to carry out the immediate activity

B. No more than the manual handling rules allow

C. Enough to last a month, but inform the site manager

D. Enough to last a week to save time going to the store

**11.16** You see a driver refuelling an excavator but most of the diesel is spilling onto the ground. What is the first thing that you should do?

A. Tell the driver immediately and locate the spill kit

B. Tell your supervisor the next time that you see them

C. Look around for a spill kit and then tell your supervisor

D. Do nothing. The diesel will eventually seep into the ground

**11.17** A worker spills a large quantity of petrol on their clothes when refuelling a piece of equipment. What should they do?

A. Put some other clothes on top

B. Change their clothes immediately

C. Nothing – it will evaporate quickly

D. Wipe it off with a cloth

**11.18** A worker spills a large quantity of petrol when refuelling a dumper. What should they do?

A. Stop – Notify – Contain

B. Stop – Contain – Notify

C. Contain – Stop – Notify

D. Notify – Contain – Stop

11

## 12    Electrical safety, tools and equipment

● Workers must be trained and competent before operating power tools.

● Cartridge-operated tools operate like a gun and can be dangerous in inexperienced hands.

● The main function of guards on cutting and grinding machines is to stop fragments flying into the air and to prevent the operator coming into contact with the blade or wheel.

● The recommended maximum voltage for construction sites is 110 volt with a yellow connector.

● Residual current devices (RCDs) should be fitted between the plug of a 230 volt tool and the supply socket.

● RCDs work by cutting the power quickly if there is a fault, and they should be mechanical (trip) tested before each use.

---

**12.01** You need to work near an electric cable but the cable has bare wires. What should you do?

A  Quickly touch the cable to see if it is live

B  Tell your supervisor and keep well away from the cable

C  Push the cable out of the way so that you can start work

D  Check there are no sparks coming from the cable and then start work

**12.02** You are using a generator to power some lighting when a lamp blows. You have a spare lamp. What should you do?

A  Carry on working in the dark

B  Replace the lamp without disconnecting the generator

C  Wait for a fully qualified electrician with a NICEIC card

D  Disconnect the lighting from the generator before replacing the lamp

---

**12.03** What are the two main visual inspections you should carry out before using a power tool?

You will be asked to 'drag and drop' your answers

A  Make sure the carry case isn't broken

B  Check it is marked with a security stamp

C  Make sure the manufacturer's label hasn't come off

D  Check the power lead, plug and casing are in good condition

E  Check switches, triggers and make sure the guards are adjusted and work correctly

**12.04** What is best practice when using a power tool with a rotating blade?

A   Adjust the guard to expose the maximum amount of blade

B   Remove the guard so that you can clearly see the blade

C   Remove the guard but wear leather gloves to protect your hands

D   Adjust the guard to expose just enough blade to let you do the job

---

**12.05** What should be done before adjusting an electric hand tool?

A   Switch it off but leave the plug in the socket

B   Put tape over the power switch before adjusting

C   Switch it off and remove the plug from the socket

D   You should never adjust an electric hand tool yourself

---

**12.06** You have been asked to use a hand tool or power tool on site. You know that it is the right tool for the job. What else must you check?

A   That it was made in the UK

B   That it is inspected before you use it

C   That it is inspected at the start of each week

D   That it was bought from a builders' merchant

---

**12.07** Why is it considered poor practice to store batteries loose in a tool bag?

A   You might forget to charge them

B   If the terminals short out, they could cause a fire

C   They give off a poisonous gas in a confined space

D   The tool bag will be heavy and could damage your back

---

**12.08** When is it safe to work close to an overhead power line?

A   If the power is switched off

B   If you use a wooden ladder for access

C   If it is not raining whilst you are working

D   If you do not touch the line for more than 30 seconds

---

**12.09** What should you do if the guard is missing from a power tool?

A   Try to make another guard

B   Use the tool but try to work quickly

C   Use the tool but work carefully and slowly

D   Do not use the tool until a proper guard has been fitted

12

**12.10** What should you do if the electrical equipment you are using cuts out?

A  Shake it to see if it will start again

B  Pull the electric cable to see if it is loose

C  Switch the power off and on a few times

D  Switch off the power and look for signs of damage

**12.11** Why should a residual current device (RCD) be used with 230 volt tools?

A  It saves energy and lowers costs

B  It lowers the voltage automatically

C  It makes the tool run at a safe speed

D  It quickly cuts off the power if there is a fault

**12.12** Which method is used to check if a residual current device (RCD) connected to a power tool is working?

A  Switch the tool on and off

B  Press the test button on the RCD

C  Use a hand-held RCD test meter

D  Run the tool at top speed to see if it cuts out

**12.13** You need to use a 230 volt item of equipment. How should you protect yourself from an electric shock?

A  Wear rubber boots and gloves

B  Put up safety screens around you

C  Use a generator which has been serviced

D  Use a portable residual current device (RCD)

**12.14** Why are battery-powered tools preferred over 110 volt tools in a construction environment?

A  They are cheaper to run

B  They are quieter

C  They are more powerful

D  They are safer

**12.15** What is the main advantage of using battery-powered tools rather than electrical ones?

A  They are cheaper to run

B  They will not give you hand-arm vibration

C  They do not need to be tested or serviced

D  They will not give you a serious electric shock

12

**12.16** Why do building sites use a 110 volt electricity supply instead of a 230 volt supply?

A   It is cheaper

B   It is less likely to kill you

C   It is safer for the environment

D   It moves faster along the cables

**12.17** What two things should you do to reduce trips and injuries caused by untidy leads and extension cables?

You will be asked to 'drag and drop' your answers

A   Only use the thinner 230 volt extension cables

B   Keep trailing cables and leads close to the wall

C   Make sure your cables have not been used before

D   Tie any excess cables and leads up into the smallest coil possible

E   Run cables and leads above head height and over the top of doorways and walkways

**12.18** What two things should you do if you need to run an electrical cable across an area used by vehicles?

You will be asked to 'drag and drop' your answers

A   Run the cable at head height

B   Cover the cable with a protective ramp

C   Cover the cable with scaffold boards

D   Put up a sign that says 'ramp ahead'

E   Wrap the cable in yellow tape so that drivers can see it

**12.19** What two things should you do if you need to use an extension cable?

You will be asked to 'drag and drop' your answers

| | |
|---|---|
| A | Uncoil the whole cable |
| B | Clean the cable with a damp cloth |
| C | Only uncoil the length of cable you need |
| D | Check the whole cable and connectors for damage |
| E | Only check the part of the cable you need for damage |

**12.20** What is the best way to protect an extension cable and also reduce trip hazards?

A  Cover the cable with yellow tape

B  Run the cable above head height

C  Run the cable by the shortest route

D  Cover the cable with pieces of wood

**12.21** What should you do if an extension cable has a cut in its outer cover?

A  Put a bigger fuse in the cable plug

B  Put electrical tape around the damaged part

C  Report the fault and make sure that no one else uses the cable

D  Check the copper wires aren't showing in the cut and then use the cable

**12.22** Apart from dust, vibration, noise and flying fragments, identify another significant hazard in this image

**12.23** What should you do if the head on your hammer comes loose?

A  Stop work and get the hammer repaired or replaced

B  Find another heavy tool to use instead of the hammer

C  Tell the other people near you to keep out of the way

D  Keep using it but be aware that the head could come off at any time

12

12.24 Which item of equipment would not require portable appliance testing (PAT)?

A. 110 volt transformer

B. Hammer and bolster

C. 110 volt extension lead

D. Plug-in breaker

12.25 Do simple hand tools like trowels, screwdrivers, saws and hammers need to be inspected?

A. No, it is not necessary to check such tools

B. Only if someone else has borrowed the tools

C. Yes, the tools should be checked each time they are used

D. Only if the tools have not been used for a few weeks

12.26 What is the main danger of using a chisel or bolster with a mushroomed head?

A. It will shatter and send fragments flying into the air

B. You are more likely to hit your hand with the chisel head

C. The hammer could slip off the head of the bolster or chisel

D. The shaft of the chisel will bend, putting a strain on your wrist

12.27 When do you need to check tools and equipment for damage?

A. Every day

B. Once a week

C. At least once a year

D. Each time before use

12.28 If a power tool has a portable appliance testing (PAT) label on it, what information should be included on the label?

A. When the tool was made

B. When the tool was last tested

C. The tool's earth-loop impedance

D. Who tested the tool before it left the factory

12.29 Which item of equipment would not require portable appliance testing (PAT)?

A. 110 volt transformer

B. Battery-powered rechargeable drill

C. 110 volt extension lead

D. Plug in portable halogen light

12

# SAFETY

**12.30** You have been asked to dig to expose power cables. You have been given a cable avoidance tool (CAT) to detect them but you haven't been shown how to use it. What should you do?

A Read the manual before you start work

B Ask a colleague to show you how to use it

C Tell your supervisor that you haven't been trained

D Dig the hole without using the cable avoidance tool

**12.31** You need to use an air-powered tool. What three hazards are likely to affect you?

You will be asked to 'drag and drop' your answers

A Radiation

B Being struck by a poorly secured hose

C An electric shock

D Hand-arm vibration

E Airborne dust and flying fragments

**12.33** What does it mean if the equipment you are using is issued with a prohibition notice?

A You must not use it until it is made safe

B You can use it as long as you take more care

C Only supervisors can use it until further notice

D You must not use it unless your supervisor is present

**12.32** If someone near you is using a petrol cut-off saw (disc cutter) to cut concrete blocks, what three immediate hazards are likely to affect you?

You will be asked to 'drag and drop' your answers

A Flying fragments

B Contact dermatitis

C Harmful dust in the air

D High noise levels

E Vibration white finger

**12.34** Which two of the following statements about power tools are true?

A You should always carry the tool by its cord

B A power tool should be unplugged by pulling its cord

C You must be trained and competent to use any power tool

D You should always unplug the tool when you are not using it

E Power tools should always be left plugged in when you check or adjust them

**12.35** Why is it dangerous to run an abrasive wheel faster than its recommended maximum speed?

A The safety guard cannot be used

B The motor could burst into flames

C The wheel will get clogged and stop

D The wheel could shatter into many pieces

12

12

# High risk activities

**13**    Site transport and lifting operations    96

**14**    Working at height    103

**15**    Excavations and confined spaces    115

**16**    Hazardous substances    119

# HIGH RISK ACTIVITIES

## 13    Site transport and lifting operations

- People being struck by moving plant is one of the most common causes of injury and death on construction sites.

- Well organised sites will have segregated vehicle and pedestrian routes.

- Vehicle marshals should be used to control, and ensure, safe vehicle movements on site.

- You must be trained, competent and authorised to operate or signal plant on any site.

- Loading and storage areas on site should be located away from main pedestrian routes.

- Poor ground conditions, excessive speed and poorly distributed loads will increase the risk of a vehicle overturning.

- One of the most common accidents involving dumpers is overturning.

- You should be provided with information about site traffic rules in your site induction.

---

**13.01** You are walking on site and a large, mobile crane reverses across your path. What **should** you do?

A   Help the driver to reverse

B   Pass close to the front of the crane

C   Wait or find another way around the crane

D   Start to run so that you can pass behind the reversing crane

**13.02** What **should** you do if you need to walk past someone operating a mobile crane?

A   Run to get past the crane quickly

B   Try to catch the attention of the crane operator

C   Take another route so that you stay clear of the crane

D   Guess what the crane operator will do next and squeeze past

**13.03** When is a site vehicle **most** likely to injure pedestrians?

A   When it is reversing

B   While digging out footings

C   While tipping into an excavation

D   As it is lifting materials onto scaffolds

**13.04** Why should you **not** walk behind a lorry when it is reversing?

A   Most lorries are not fitted with mirrors

B   The driver is unlikely to know you are there

C   Most lorry drivers aren't very good at reversing

D   You will need to run, not walk, to get past it in time

**13.05** The quickest way to your work area is through a contractor's vehicle compound. Which way **should** you go?

A | Around the compound every time

B | Around the compound if vehicles are moving

C | Straight through the compound if no-one is looking

D | Straight through the compound if no vehicles appear to be moving

**13.06** When is site transport allowed to drive along a pedestrian route?

A | During meal breaks

B | If it is the shortest route available

C | Only if the vehicle has a flashing yellow light

D | Only if necessary and if all pedestrians are excluded

**13.07** When you walk across the site, what is the **best** way to avoid an accident with mobile plant?

A | Ride on the plant

B | Wear hi-vis clothing

C | Keep to the designated pedestrian routes

D | Get the attention of the driver before you get too close

**13.08** You need to walk past a 360° mobile crane. The crane is operating near a wall. What is the **main** danger?

A | The crane could crash into the wall

B | You could get whole-body vibration from the crane

C | You could be crushed if you walk between the crane and the wall

D | Your hearing could be damaged by high noise levels from the crane

**13.09** A forklift truck is blocking the route you need to take on site. It is lifting materials onto a scaffold. What **should** you do?

A | Start to run so that you are not under the load for very long

B | Wait or take another route, but never walk under a raised load

C | Catch the driver's attention and then walk under the raised load

D | Only walk under the raised load if you are wearing a safety helmet

**13.10** Which action **should** a worker take if they see mobile plant using a route intended only for pedestrians?

A | Nothing, the driver will know what they are doing

B | Report this to their supervisor

C | Have a word with the operator at the end of the day

D | Just be careful in that area

13

**13.11** Workers are on foot close to moving plant. Which **one** of the following is **true**?

A. Hi-vis clothing will keep the workers safe if the plant is not reversing

B. The operator will see the workers, because they have mirrors and CCTV

C. The workers should stay within the designated pedestrian routes

D. The workers will be safe if they are in a group

**13.12** Which of the following signs means **No pedestrian access**?

A.

B.

C.

D.

**13.13** In which one of the following situations is it **safer** for a worker to speak to someone operating mobile plant?

A. The operator knows the worker is there and the plant has stopped operating

B. The worker is wearing hi-vis and the plant is moving slowly

C. The operator can hear the worker and it is daytime

D. The worker is wearing PPE and the plant is moving slowly

**13.14** What **should** you do if you see a dumper being driven too fast?

A. Report it to the police

B. Keep out of its way and report it

C. Try to catch the dumper and speak to the driver

D. Nothing, as dumpers are allowed to speed

**13.15** What is the **main** hazard associated with the movement of plant and machinery around site?

A. Pedestrians walking too close to moving machinery and being crushed

B. Existing building collapse from vibrations of the moving machinery

C. Members of the public being poisoned by the exhaust fumes

D. Drivers getting motion sickness from the movement of the machine

**13.16** What is the meaning of this sign?

A. Pedestrian walkway only

B. No pedestrian access

C. Traffic approaching from each direction

D. Go slow

13

**13.17** When moving plant or machinery around site, what should the operator look out for?

- A Driving with the hand-brake on
- B Driving with the lights on during the day
- C Speed signs and speed humps
- D Only driving with limited fuel

**13.18** Where vehicles tip materials into excavations, what **could** be used as a safety precaution?

- A Stop blocks
- B Extra speakers
- C Flashing lights
- D A siren

**13.20** If there are blind spots while using plant but work needs to continue, what actions **should** be taken?

- A Use the existing mirrors on the plant
- B Request the plant be fitted with CCTV cameras
- C Use a vehicle marshal for this type of work
- D Work with a slinger

**13.21** Your supervisor asks you to drive a dumper truck but you have **not** driven one before. What **should** you do?

- A Ask a trained driver how to operate it safely
- B Watch other dumpers to see how they are operated
- C Operate the dumper in an open area in case you make a mistake
- D Tell your supervisor that you are not trained and so cannot operate it

**13.19** Why **should** engines be turned off before leaving a site vehicle? Select **two** answers.

You will be asked to 'drag and drop' your answers

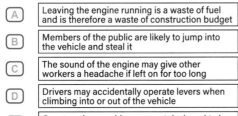

- A Leaving the engine running is a waste of fuel and is therefore a waste of construction budget
- B Members of the public are likely to jump into the vehicle and steal it
- C The sound of the engine may give other workers a headache if left on for too long
- D Drivers may accidentally operate levers when climbing into or out of the vehicle
- E Construction machines are not designed to be constantly left running

13

# HIGH RISK ACTIVITIES

**13.22** Which of the following is a **recognised** control measure when reversing a vehicle?

A    Turning the site radio off

B    Using a vehicle marshal

C    Turning on all the vehicle lights

D    Standing on the back to direct it

**13.23** You think a load is about to fall from a moving forklift truck. What **should** you do?

A    Run and tell your supervisor

B    Sound the nearest fire alarm bell

C    Run alongside the machine and try to hold onto the load

D    Keep clear but try to warn the driver and others in the area

**13.24** When can a mobile plant operator let people ride in, or on, the machine?

A    Only if they have a long way to walk

B    Any time as long as the cab door is shut

C    Any time as long as the site speed limit is not exceeded

D    Only if it is designed to carry passengers and has a designated seat

**13.25** You see a lorry parking and it has a flat tyre. Why **should** you tell the driver?

A    More fuel will be used by the lorry

B    It could be unsafe to drive the lorry

C    The lorry can only carry small loads

D    The driver will need to travel at a much slower speed

**13.26** An excavator has just stopped work. Liquid is dripping and forming a small pool under the back of the machine. What could this mean?

A    The machine has a leak and could be unsafe

B    It is normal for fluids to vent after the machine stops

C    The machine is hot so the diesel has expanded and overflowed

D    Someone put too much diesel into the machine before it started work

**13.27** A mobile crane is lifting a load but the load is about to hit something. What **should** you do?

A    Go and tell your supervisor

B    Go and tell the crane driver

C    Try and warn the person supervising or signalling the crane

D    Do nothing and assume that everything is under control

13

**13.28** Which signal is being shown in this image?

| | |
|---|---|
| A | Danger, emergency stop |
| B | Move forwards |
| C | Turn right or left |
| D | Move backwards |

---

**13.29** What is needed before supervising any lift using a crane?

| | |
|---|---|
| A | A mobile phone to talk to the crane driver |
| B | Full training and being assessed as competent |
| C | Written instructions from the crane hire company |
| D | Nothing. The crane driver will tell you what to do |

**13.30** Which of the following is a way of ensuring that a slinger or signaller is trained and competent?

| | |
|---|---|
| A | By trusting them when they say they are |
| B | By asking for evidence to be produced |
| C | By having them swear an affidavit |
| D | By making a handshake agreement |

**13.31** Which action will help to keep signallers safe?

| | |
|---|---|
| A | Provide yearly eye tests to confirm they have good vision |
| B | Provide body cameras to capture any incidents |
| C | Provide gloves for hand signals |
| D | Provide hi-vis clothing so they are clearly visible |

**13.32** Under what circumstance **should** a driver stop their vehicle immediately?

| | |
|---|---|
| A | If the vehicle is low on fuel |
| B | If the flashing beacon has stopped working |
| C | If they lose sight of their vehicle marshal |
| D | If they are operating in a one-way system |

13

**13.33** When signallers are used, who **should** they be in contact with at all times?

A. The machine operator

B. The site manager

C. Their supervisor

D. Pedestrians

**13.34** What is the most important information a vehicle marshaller should know before directing a vehicle?

A. How to signal vehicles and any relevant safety procedures

B. The type of materials being delivered to the site

C. The name and address of the driver for security reasons

D. The value of the materials, as they could be stolen

**13.35** To reduce the risk of overturning and accidents when not in use, how **should** earth moving vehicles be parked?

A. With their buckets and blades raised in the air

B. With their buckets and blades facing the same way

C. With their buckets and blades lowered to the ground

D. With the buckets and blades facing opposite directions

**13.36** Where risk of overturning is significant, what **should** vehicles be fitted with?

A. Extra strength brakes

B. Roll-over protective structures (ROPS)

C. Heavy duty graded tyres

D. A winch and pulley system

**13.37** To prevent over-turning, when should rear tipping lorries **not** be used for tipping operations?

A. When on firm, level ground

B. On uneven or sloping ground

C. When a competent signaller is supervising

D. During redistribution of the load

13

## 14    Working at height

- Work at height is defined as work at any height where a person could fall and be injured.

- Every year falls from height kill more construction workers than any other type of accident.

- Work at height should be avoided where possible.

- If you are involved in work at height, your employer should ensure that you have sufficient information, instruction and equipment so that you can work safely.

- All equipment for working at height should be inspected before use.

- There should always be a rescue plan if people are working at height.

- If you feel that the task you are completing at height is unsafe, stop work and report it to your supervisor.

- It would be classed as working at height if you were standing on the back of a lorry during loading or unloading activities.

- All roofs should be treated as fragile until a competent person has confirmed they are not.

*Fragile roof*

- Safe access and a safe working platform should be provided for all work on fragile roofs.

- One of the leading causes of injury on construction sites is as a result of workers being struck by falling objects.

- Do not attempt to erect, alter or dismantle a mobile access tower unless you have been trained and you are authorised to do so.

14

- Make sure that there are no people, tools or equipment on a mobile access tower before you attempt to move it.

- The erection, alteration, inspection and dismantling of scaffolding should only be carried out by trained and authorised persons.

- Personal fall-arrest equipment is designed to minimise the consequences if a fall occurs, and will only protect an individual worker.

# HIGH RISK ACTIVITIES

**14.01** Where **should** vehicles be loaded and unloaded?

A  On an upward slope

B  On level ground

C  On a downward slope

D  On uneven ground

**14.02** What is the purpose of a one-way system at a loading or unloading area?

A  To eliminate the need to reverse

B  To allow faster speed limits

C  To reduce the speed limits

D  To increase the need to reverse

**14.03** Accidents on site are often caused by materials falling from vehicles during what process?

A  Refuelling

B  Repainting

C  Cleaning

D  Unloading

**14.04** What **should** you do if you find a ladder that is damaged?

A  Try to mend the damage before using it

B  Use it if you can avoid the damaged part

C  Don't use it and report the damage at the end of your shift

D  Don't use it and make sure that others know about the damage

**14.05** Which image shows the **safest** method for using a stepladder?

A

B

C

D

14

**14.06** Which image shows the **safe** use of a ladder?

(A)

(C)

(B)

(D)

---

**14.07** What angle should a leaning ladder be used at?

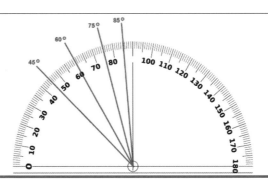

**14.08** According to the Work at Height Regulations, when **can** ladders be used for work?

 (A) If it is high risk work for long periods of time

 (B) A ladder must never be used on site

(C) If it is low risk work for a short period of time

 (D) When other people do not need to use it for access

**14.09** Who **should** check a ladder before it is used?

(A) The site manager

 (B) The manufacturer

 (C) A site safety officer

 (D) The person who is going to use it

14

**14.10** What is the **best** way to make sure that a ladder is secure and will not slip?

A  Secure it at the top

B  Secure it at the bottom

C  Wedge the bottom of the ladder with blocks of wood

D  Ask someone to stand with their foot on the bottom rung

**14.11** What is the **correct** way to climb a ladder?

A  By having two people on the ladder at all times

B  Only using the ladder when wearing a safety harness

C  Having two points of contact with the ladder at all times

D  Having three points of contact with the ladder at all times

**14.12** How many people are allowed on a ladder at the same time?

A  Only one person

B  A maximum of two people

C  Three people, if it is long enough

D  One person on each section of an extension ladder

**14.13** Which of the following is **not** true when using podium steps?

A  Podium steps are safe and can't topple over

B  Podium steps should be inspected before use

C  Podium step wheels must be locked before you get on

D  Podium steps can easily topple if you overreach sideways

**14.14** Which of these statements is **true** about using a ladder to access a scaffold platform?

A  All broken rungs must be clearly marked

B  Two people must be on the ladder at all times

C  It must be wedged at the bottom to stop it slipping

D  It must be secured, and extend at least 1 m above the platform

**14.15** What should you do if you need to use a mobile access tower but the brakes don't work?

A  Do not use the tower

B  Only use the tower if the floor is level

C  Get someone to hold the tower while you use it

D  Use some wood to wedge the wheels and stop them moving

14

**14.16** What is the **correct** way to reach the working platform of a mobile access tower?

A Climb up the ladder built into the tower

B Climb up the outside of the diagonal bracing

C Lean a ladder against the tower and climb up that

D Climb up the tower frame on the outside of the tower

**14.17** A mobile access tower must **not** be used on what surface?

A A paved patio

B An asphalt road

C Soft or uneven ground

D A smooth concrete path

**14.18** Which of the following is the **safest** method of accessing a mobile access tower?

A Climbing up the outside of the tower

B Climbing a ladder inside the tower

C Climbing a ladder outside of the tower

D Climbing a rope on the outside of the tower

**14.19** Which **one** of the following statements is **true** when referring to the wheels on mobile access towers?

A The wheels should be locked at all times

B The wheels should be locked when the tower is in use

C The wheels should be locked when the tower is being moved

D The wheels should only be locked at the end of the day

**14.20** Which **one** of the following is a safe way of moving a mobile access tower?

A Towing with a site vehicle, with a tow rope attached to the base

B Using manual effort pushing only from the base

C Using manual effort to pull from the top and the base

D Towing with a site vehicle with a tow rope attached to the top

**14.21** When assembling a mobile access tower, near overhead electric cables, which **one** of the following statements is **true**?

A The cables can be treated as dead if the work is going to take less than 30 minutes

B The cables must be treated as live until it is confirmed they are dead

C The cables do not present a danger because mobile access towers are insulated

D Personal protective equipment (PPE) will keep workers safe until it is confirmed that the cables are dead

14

**14.22** When working at height in a mobile elevating work platform (MEWP), over or near to deep water, which item of personal protective equipment (PPE) **should** be worn?

A — Wellington boots

B — Lifejacket

C — Full face respirator

D — Full body harness

**14.23** What should a harness's lanyard be attached to when working in a mobile elevating work platform (MEWP)?

A — The control box

B — The MEWP handrail

C — A point on the structure or building you are working on

D — A designated anchor point within the platform or basket

**14.24** A worker has been asked to operate a mobile elevating work platform (MEWP), but has no training. What **should** they do?

A — Get the work done as quickly as possible

B — Ask a workmate how to operate the MEWP

C — Tell their supervisor that they have no training

D — Operate the MEWP at breaktime when no-one is around

**14.25** A mobile elevating work platform (MEWP) must **not** be used on what surface?

A — An asphalt road

B — A smooth concrete path

C — Soft or uneven ground

D — A concrete road

**14.26** A worker is wearing a harness in a boom type mobile elevating work platform (MEWP) (sometimes known as a cherry picker). Which **one** of the following **should** the harness have?

A — A work-restraint lanyard clipped to an attachment point in the basket

B — A work-restraint lanyard clipped to the hand rail of the basket

C — A fall-arrest lanyard clipped to the structure being worked on

D — A fall-restraint lanyard clipped to the control box of the machine

**14.27** What does this symbol on a mobile elevating work platform (MEWP) show?

A — The location of the lowering controls for use in emergency

B — The location of the points where workers can lean over the platform

C — The guard-rail height

D — The safe method of exiting the platform

**14.28** Which **one** of the following should a worker do if a mobile elevating work platform (MEWP) does **not** allow safe access to the place of work?

A   Stand on the guard rails

B   Use a step ladder on the platform

C   Inform a supervisor that a larger MEWP is needed

D   Put pallets on the working platform

**14.29** What should you do if you are required to use access equipment that you have **not** been trained to use?

A   Get a ladder instead

B   Ask someone else to do it

C   Do the job if it won't take long

D   Stop work and speak to your supervisor

**14.30** If you are working on a flat roof, what is the **best** way to stop yourself falling over the edge?

A   Use red and white tape to mark the edge

B   Put a large warning sign at the edge of the roof

C   Protect the edge with a guard-rail and a toe-board

D   Ask someone to watch you and shout when you get too close to the edge

**14.31** Who **should** complete pre-use checks on ladders or other equipment used for working at height?

A   The employer

B   The worker using the equipment

C   The supervisor

D   The site manager

**14.32** Which **two** of the following statements are **true** about working on a roof?

You will be asked to 'drag and drop' your answers

A   It is safe to try to walk near underlying roof supports

B   Wired glass roof panels are likely to be fragile

C   It is easy to see which roof surfaces are likely to be fragile

D   Workers should not work on a roof where there is no protection from falls

E   Asbestos and fibre cement roof sheets are unlikely to be fragile

# HIGH RISK ACTIVITIES

**14.33** What does the following sign mean?

A Fragile roof

B Deep water

C Safety boots must be worn

D No running

**14.34** A fragile roof needs to be repaired. Which of the following would be regarded as the **safest** method?

A Workers work from underneath using a mobile elevating working platform (MEWP)

B Workers access the roof by walking as close as possible to the underlying roof supports

C Workers working on the roof wearing safety boots and helmets

D Workers work from underneath using ladders and ropes for anchoring

**14.35** Which **one** of the following statements is **true** of a person who has fallen, and is suspended in a fall-arrest harness?

A They will need to be rescued quickly

B There will be no reason to call an ambulance

C They will be safe in the harness for over an hour

D They should be left to rescue themselves

**14.36** Which **one** of the following statements is **true** of painting wooden ladders used in construction?

A Ladders should be painted orange to make them more visible

B Ladders should never be painted as this could hide defects or damage

C It is a good idea to paint ladders because this protects them

D It is advisable to paint ladders to prevent them being stolen

**14.37** A worker is storing materials above toe-board height on a scaffold. How **should** people below be protected?

A Shout a warning to anyone passing below

B Use string to secure the materials

C Halt the work when people are approaching

D Use a brick guard or suitable mesh netting

**14.38** Under which **one** of the following circumstances is it safe for a worker to remove a protective cover from a deep service hole on site?

A The worker is wearing a safety helmet and waits until everyone else has left the site

B The worker is authorised to do so and is protected from falling whilst the cover is not in place

C The worker is wearing hi-vis and has told a few people on site that the cover will be removed

D The worker has placed a safety cone by the hole so people will avoid the area

14

**14.39** What is the **maximum** length of time that a worker should work from a step ladder in one position?

A   Less than 30 minutes

B   Less than one hour

C   Less than 90 minutes

D   Less than two hours

**14.40** Which **one** of the statements about storing materials on a working platform is **correct**?

A   Materials can be stored unsecured but they must be above guard-rail height

B   Materials do not need to be secured if they are going to be there for less than an hour

C   Materials can be stored anywhere, even if they pose a trip hazard or block the walkway

D   Materials must be stored so they can't fall and the platform must be able to take their weight

**14.41** What is the **best** way to stop people being hit by falling tools and materials when you are working above them?

A   Make sure they are wearing safety helmets

B   Only allow authorised people underneath the work area

C   Tell them you will be working above them and erect signs

D   Exclude people from below the work area with fencing and signs

**14.42** If you need to stack materials on a working platform, what is the **best** way to stop them falling over the toe-board?

A   Cover the stack with polythene

B   Put a warning sign on the stack

C   Have brick guards or netting fitted to the edge

D   Build the stack so that it leans away from the edge

**14.43** To ensure the public is not put at risk from falling materials for the duration of work, what may be necessary?

A   Pavement closure or diversion

B   Giving the public hard hats

C   Making pedestrians use the road

D   Giving the public hi-vis clothing

**14.44** If a person is struck by a falling object, what **could** be the negative consequence?

A   They get fired

B   They get injured

C   They get a promotion

D   They get compensation

14

**14.45** What piece of personal protective equipment (PPE) **should** be worn on sites where there is a risk of falling objects?

A   Protective goggles

B   Hard hat

C   Hi-vis clothing

D   A safety harness

**14.46** When is it safe to cross a fragile roof?

A   Only when you can see fragile roof signs

B   Only if you do not walk on any plastic panels

C   When crawling boards with handrails are available to use

D   At any time as long as you walk along the line of the bolts

**14.47** What does this sign mean?

A   Fragile roof. Take care when walking on roof surface

B   Load-bearing roof. The surface can be slippery when wet

C   Load-bearing roof. You can stand on the surface but not on any roof lights

D   Fragile roof. Use fall protection measures and do not stand directly on the roof

**14.48** What is the **best** way to stop people falling through voids, holes or fragile roof panels?

A   Tell everyone where the dangerous areas are

B   Mark the dangerous areas with red and white warning tape

C   Cover the dangerous areas with safety netting and tell everyone to be careful

D   Place secure, load-bearing covers over the dangerous areas and add warning signage

**14.49** A material that may hide fragile surfaces has been applied to a roof. What action **should** be taken?

A   Nothing – the material applied should be fine

B   The fragile areas should be clearly marked and protected

C   Nothing – workers should know to be careful

D   The fragile areas should be painted green

**14.50** Which **one** of the following surfaces is **not** likely to be fragile?

A   A reinforced concrete roof

B   A fibre cement sheet roof

C   A glass panel roof

D   A slate tiled roof

**14.51** A scaffold guard-rail **must** be removed to allow you to carry out a task. If you are not a scaffolder, can you remove the guard-rail?

A ) Yes, if you put it back before you leave the site

B ) Yes, if you put it back as soon as you have finished

C ) No, only a scaffolder can remove the guard-rail and put it back

D ) No, only a scaffolder can remove the guard-rail but you can put it back

**14.52** How can the safe load rating for a scaffold platform be identified?

A ) By asking the telehandler driver

B ) By asking the principal contractor

C ) Referring to the handover certificate or signage

D ) The safe load is breached when the ledgers start to deflect

**14.53** What should you do if you think that the scaffolding you are working from is **not** safe?

A ) Report it to your supervisor at the end of the shift

B ) Try to make the repairs yourself and then report it to your supervisor

C ) Report your concerns to your supervisor straight away

D ) Ignore it and wait for the scaffolders to identify any problems

**14.54** What should you do if you notice your harness or attachment is damaged?

A ) Use a colleague's harness instead

B ) Stop and tell your supervisor straight away

C ) Use it and tell your supervisor at the end of the day

D ) Stop and tell your supervisor but carry on using it until it is replaced

**14.55** What is an **inertia reel**?

A ) A retractable fall arrester

B ) A horizontal fall arrester

C ) A pulley-operated fall arrester

D ) A rope-based fall arrester

**14.56** Which of the following **best** describes the purpose of personal fall prevention equipment?

A ) It is designed to prevent falls from occuring

B ) It is designed to minimise the consequences if a fall occurs

C ) It is designed to protect more than one person if a fall occurs

D ) It is designed to be used in confined spaces only

14

**14.57** When are personal fall-arrest systems to be used?

A Only as a last resort

B The majority of the time

C In the morning

D During a night shift

**14.58** Who **should** know how to carry out pre-use checks on fall-arrest equipment?

A The site managers

B All workers who use it

C All workers on site

D The apprentice workers

14

## 15    Excavations and confined spaces

- Excavations should always have a safe means of access and egress, such as a secured ladder.
- Excavations should be inspected at the start of every shift, or after events that might affect stability.
- The most accurate way to identify the location of buried services is through the use of trial holes.
- If you damage an underground service, stop work, do not touch anything and report it.
- Permit systems are often used where people are working in confined spaces.
- If you are working in a confined space and the gas alarm sounds, get out immediately.
- There should always be a rescue plan if you are working in a confined space.

---

**15.01** You are in a deep trench. A lorry backs up to the trench and the engine is left running. What **should** you do?

A  Get out of the trench quickly

B  See if there is a toxic gas meter in the trench

C  Put on ear defenders to cut out the engine noise

D  Ignore the problem, as the lorry will soon drive away

**15.02** What **should** you do if you see the side supports move when you are working in an excavation?

A  Work in another part of the excavation instead

B  Keep working and watch to see if they move again

C  Make sure that you and other workers get out quickly

D  Nothing. The sides are expected to move all the time

**15.03** What is the **main** hazard when working in an excavation?

A  Breathing-in hazardous dust from the earth

B  Cuts and abrasions from the trench sides

C  Trips and falls due to the space restriction

D  Crushing, if the sides are not supported

**15.04** When **should** an excavation be battered back or stepped?

A  If it is more than 5 m deep

B  If any buried services cross the excavation

C  If there is water in the bottom of the trench

D  If there is a risk of the sides falling in, regardless of depth

15

# HIGH RISK ACTIVITIES

**15.05** What do guard-rails around the top of an excavation prevent?

- **A** The sides of the trench collapsing
- **B** People falling into the trench and being injured
- **C** Toxic gases collecting in the bottom of the trench
- **D** Rainwater running off the ground at the top and into the trench

**15.06** What is the **safest** way to get into and out of a deep excavation?

- **A** Use a fixed staircase
- **B** Use an excavator bucket
- **C** Use the buried services as steps
- **D** Use the shoring or trench supports

**15.07** What equipment **should** be used when digging near to underground electrical services?

- **A** An excavator
- **B** A jack hammer
- **C** A pick and fork
- **D** An insulated spade

**15.08** What does it mean if a run of coloured marker tape is found when digging?

- **A** The excavation now requires side supports
- **B** There are buried human remains and you must tell your supervisor
- **C** There is a buried service and further excavation must be carried out with care
- **D** The soil is contaminated and you must wear respiratory protective equipment (RPE)

**15.09** According to the guidance on underground service pipes, what does a yellow service pipe carry?

- **A** Water
- **B** Gas
- **C** Electricity
- **D** Telecoms

15

**15.10** What **three** things should you do before entering a confined space that has sludge at the bottom?

You will be asked to 'drag and drop' your answers

A Identify what the sludge is

B Have the correct training

C Put on a disposable dust mask

D Make sure that the space has been tested for gas

E Throw something into the sludge to see how deep it is

**15.11** What **should** you do if your permit to work in a confined space will run out before you finish the task you are carrying out?

A Hand the permit over to the next shift

B Carry on working until the job is finished

C Leave the confined space before the permit runs out

D Ask your supervisor to change the date on the permit

**15.12** What **should** you do if you are in a deep trench and you start to feel dizzy?

A Sit down in the trench and take a rest

B Get out, let your head clear and then go back in again

C Carry on working and hope that the feeling will go away

D Make sure that you and any others get out quickly and report it

**15.13** Why is methane gas dangerous in confined spaces? Give **two** answers

You will be asked to 'drag and drop' your answers

A It can explode

B It makes you hyperactive

C It makes you dehydrated

D You may not have enough oxygen to breathe

E You will not be able to see because of the dense fumes

15

# HIGH RISK ACTIVITIES

**15.14** What is the **most** important reason why people should be trained and competent before they are allowed to enter a confined space?

A. Confined spaces never contain breathable air

B. Confined spaces are only found on house-building sites

C. Confined spaces always contain flammable or explosive gases

D. Confined space entrants need to understand the potential hazards

**15.15** What is the **main** reason for having a person positioned immediately outside a confined space while work is taking place inside it?

A. To carry out a risk assessment for the work

B. To check compliance with the method statement

C. To start the rescue plan if there is an emergency

D. To supervise the work inside the confined space

**15.16** What is the **main** cause of people dying while working in a confined space?

A. Lack of oxygen

B. Too much oxygen

C. Presence of methane

D. Cold conditions leading to hypothermia

**15.17** What might happen if the level of oxygen drops below 8% in a confined space?

A. You might get dehydrated

B. Your hearing could be affected

C. You could become unconscious

D. There is a high risk of fire or explosion

**15.18** When working in a confined space, what is it a sign of if there is a smell of rotten eggs?

A. Oxygen

B. Methane

C. Carbon dioxide

D. Hydrogen sulphide

**15.19** What is it likely to mean if the soil gives off a strange smell when digging?

A. The soil contains a lot of clay

B. The ground could be contaminated

C. The soil has been excavated before

D. The ground has been used to grow crops in the past

15

# 16 Hazardous substances

- Asbestos containing materials (ACMs) can be difficult to identify. Asbestos is made up of hazardous, microscopic fibres which can easily be inhaled.

- If you think a material contains asbestos, always assume it does. Stop work, warn others, and report it to your supervisor.

- Your employer should ensure that exposure to hazardous substances is prevented or adequately controlled.

- Health and safety information for hazardous substances should be detailed in a COSHH assessment.

- Control measures for working with hazardous substances should be monitored regularly.

- Wet cement and concrete can cause skin burns and dermatitis if they are in direct contact with your skin.

- Lead is toxic. The most common route of entry into the body is via the mouth (ingestion).

**16.01** Where are you **most** likely to come across asbestos?

- A  In a house built between 1950 and 2005
- B  In any industrial building built after the year 2000
- C  In any building built or refurbished before the year 2000
- D  Asbestos has now been removed from all houses and buildings

**16.02** Breathing in asbestos dust is **most** likely to cause which of the following?

- A  Lung diseases
- B  Throat infections
- C  Dizziness and headaches
- D  Aching muscles and painful joints

16

**16.03** Exposure to asbestos fibres may result in which illness?

A Dermatitis

B Skin cancer

C Heart disease

D Lung cancer

**16.04** How can asbestos be correctly identified?

A The distinct colour of the dust

B By getting a sample analysed in a laboratory

C It is clear from the strong smell of the dust

D By putting a piece in water and seeing if it dissolves

**16.05** Which one of the following statements about asbestos is true?

A Asbestos fibres are most likely to enter the body through the skin

B Asbestos fibres only cause health problems for smokers

C Asbestos in buildings must always be removed regardless of condition

D Asbestos fibres are most likely to enter the body through inhalation

**16.06** Cement-based roofing sheets are a common material which can often contain what hazardous substance?

A Rust

B Dry rot

C Termites

D Asbestos

**16.07** What is the main, immediate hazard from kneeling directly on wet cement?

A Skin burns

B Dermatitis

C Eczema

D Skin rash

**16.08** What does a COSHH assessment cover?

A Working safely in confined spaces

B Lifting heavy loads and how to protect yourself

C The assessment of noise levels and how to protect your hearing

D Hazardous substances and how to protect yourself when using them

16

**16.09** Whose responsibility is it to explain the health risks and safe method of work you need to follow (the COSHH assessment) before work starts with a hazardous substance?

A  The site first aider

B  The site security people

C  Your supervisor or employer

D  A Health and Safety Executive (HSE) inspector

**16.10** What is the **first** thing you **should** do if you find an unmarked container that you think might contain chemicals?

A  Smell it to see what the chemical is

B  Move the container to somewhere safe

C  Put the container in a bin to get rid of it

D  Ensure that it remains undisturbed and report it

**16.11** How is it possible to tell that a product is hazardous?

A  It will always be in a cardboard box

B  It will always be in a black container

C  By the shape of the container or packaging

D  By warning symbols on the container or packaging

**16.12** What does the word **sensitiser** mean on the packaging of a substance?

A  It should not be used under any circumstances

B  It must be mixed with water before it can be used

C  That it could cause allergic reactions when handled

D  It is safe to use without personal protective equipment (PPE)

**16.13** Identify which of the following signs is associated with a substance being toxic if swallowed or inhaled?

A

B

C

D

**16.14** If warnings about how to work with hazardous substances are not followed, what is a **likely** consequence for workers?

A  Good health

B  Increased fitness level

C  Decreased fitness levels

D  Ill health

16

# HIGH RISK ACTIVITIES

**16.15** What should employers check regularly if you are working with hazardous substances?

A. Your mood

B. Your family

C. Your health

D. Your wages

**16.16** A worker is using a new substance when they start to feel ill. What should the worker do?

A. Stop work and report it to a supervisor or manager on site immediately

B. Nothing – it is acceptable to feel ill with certain substances

C. Continue with the work but report it to the supervisor later

D. Enter the details into an incident report and continue to work with the substance

**16.17** When working through a construction health and safety checklist, which of these hazardous substances **should** be identified?

A. Lead, solvents, cement, asbestos

B. Asbestos, cement, paints, noise levels

C. Noise levels, solvents, dust, paint

D. Vibration levels, noise levels, asbestos, cement

**16.18** Which of the following tasks could place a worker at the **greatest** risk of lead poisoning, if control measures were not put in place?

A. Plastering a ceiling in a new build home

B. Building a wall out of old stone

C. Cutting timber in a roof construction

D. Sanding down some old paintwork

16

# Congratulations

## You have now completed the core knowledge questions

| For the Specialists test |
| --- |
| You should now revise the appropriate specialist activity from Section E. |

# Specialist

If you are preparing for a specialist test you also need to revise the appropriate specialist activity, from those listed below.

| | | |
|---|---|---|
| **17** | Supervisory | 126 |
| **18** | Demolition | 140 |
| **19** | Highway works | 147 |
| **20** | Specialist work at height | 156 |
| **21** | Lifts and escalators | 165 |
| **22** | Tunnelling | 174 |

Heating, ventilation, air conditioning and refrigeration (HVACR)

| | | |
|---|---|---|
| **23** | Heating and plumbing services | 182 |
| **24** | Pipefitting and welding | 190 |
| **25** | Ductwork | 199 |
| **26** | Refrigeration and air conditioning | 207 |
| **27** | Services and facilities maintenance | 215 |

| | | |
|---|---|---|
| **28** | Plumbing (JIB) | 223 |

**17.01** What is the purpose of the health and safety file that is handed to the client at the end of the project?

A   To help people who have to carry out work on the structure in the future

B   To help prepare the final accounts for the structure

C   To record the health and safety standards of the structure

D   To record the accident statistics of the construction project

**17.02** When the Construction (Design and Management) Regulations 2015 apply, what **must** be in place before construction work begins?

A   The health and safety file

B   The construction phase plan

C   The method statement

D   The construction contract agreement

**17.03** Under the Construction (Design and Management) Regulations 2015, what **must** be provided before construction work starts, and then maintained until the end of the project?

A   A safety log book

B   A premises log book

C   A car park or other parking facilities

D   Adequate welfare facilities

**17.04** Under the Construction (Design and Management) Regulations 2015, where would you find the arrangements for managing health and safety for the project you are working on?

A   In the health and safety file

B   In the construction phase plan

C   In the contract documentation

D   In the designer's risk assessment

**17.05** Under the Construction (Design and Management) Regulations 2015, which of the following **must** be in place before demolition work can start?

A   A health and safety file

B   The arrangements for demolition recorded in writing

C   A demolition risk assessment

D   The pre-tender demolition health and safety plan

**17.06** Under the Construction (Design and Management) Regulations 2015, when the contractor sets a person to work on a construction site, what **must** they ensure that person has, or be in the process of, obtaining?

A   The right skills, knowledge, training and experience

B   The relevant competency card

C   A hard hat, hi-vis clothing and safety footwear

D   A relevant qualification for the work to be undertaken

**17.07** Under the Construction (Design and Management) Regulations 2015, which **two** of the following must you ensure workers have received before they start working on site?

[A] A suitable site induction, specific to the work

[B] Details of the client's brief and project expectations

[C] Confirmation of their working hours and rest breaks

[D] Details of the designer's plan of work

[E] Information on relevant hazards and control measures

**17.08** How long **must** you keep inspection records under the Construction (Design and Management) Regulations 2015?

[A] For three months after the inspection has been carried out

[B] For one week on site before sending them to head office

[C] Until the construction work is complete and then for three months

[D] Only until the project is complete

**17.10** What does a COSHH assessment tell you?

[A] How to lift heavy loads and how to protect yourself

[B] How to work safely in confined spaces

[C] How to use a substance safely in the environment in which it is to be used

[D] How to assess noise levels to protect your hearing

**17.11** Which piece of equipment is used with a cable avoidance tool (CAT) to detect cables?

[A] Compressor

[B] Signal generator

[C] Metal detector

[D] Gas detector

**17.09** You have to use a new substance for the first time and need to carry out a COSHH assessment. What are the **two** main things you will need?

You will be asked to 'drag and drop' your answers

[A] Your company's safety policy

[B] The safety data sheet

[C] The age of the people doing the work

[D] The delivery note

[E] Details of where, and how, you intend to use the substance

17

**17.12** On the site electrical distribution system, which colour plug indicates a 400 volt supply?

A  Yellow

B  Blue

C  Black

D  Red

**17.13** Why **must** a residual current device (RCD) be used in conjunction with 230 volt electrical equipment?

A  It lowers the voltage

B  It quickly cuts off the power if there is a fault

C  It makes the tool run at a safe speed

D  It saves energy and lowers costs

**17.14** How could a site worker check if the residual current device (RCD) through which a 230 volt hand tool is connected to the supply is working correctly?

A  Switch the tool on and off

B  Press the test button on the RCD unit

C  Switch the power on and off

D  Run the tool at top speed to see if it cuts out

**17.15** In the colour coding of electrical power supplies on site, what voltage does a blue plug represent?

A  50 volts

B  110 volts

C  230 volts

D  400 volts

**17.16** Where should an emergency escape route take you to?

A  The ground

B  The open air

C  A place of safety

D  A first-aid room

**17.17** Which of the following is a significant hazard when excavating alongside a building or structure?

A  Undermining or weakening the foundations of the building

B  Noise and vibration affecting the occupiers of the building

C  Excavating too deeply into soft ground

D  Damaging the surface finish of the building or structure

**17.18** What danger is created by excessive oxygen in a confined space?

A An increase in the breathing rate of workers

B An increased flammability of combustible materials

C An increased working time inside the work area

D A false sense of security

**17.19** When planning possible work in a confined space, what should be the first consideration?

A How long the job will take

B How to avoid the need for operatives to enter the space

C How many operatives will be required

D What personal protective equipment (PPE) will be needed

**17.20** When is it advisable to take precautions to prevent people, plant or materials falling into an excavation?

A At all times

B When the excavation is 2 m or more deep

C When the excavation is 1.2 m or more deep

D When there is a risk from an underground cable or other service

**17.21** Which of the following precautions should be taken to prevent a dumper that is tipping material into an excavation from falling into it?

A Dumpers should be kept 5 m away from the excavation

B Stop blocks should be provided, parallel to the trench and appropriate to the vehicle's wheel size

C Dumper drivers are required to judge the distance carefully or be given stop signals by another person

D Cones or signage should be erected to indicate the safe tipping point

**17.22** Which two of the following factors must be considered when providing first-aid facilities on site?

You will be asked to 'drag and drop' your answers

A The hazards, risks and nature of the work carried out

B The number of people expected to be on site at any one time

C The difficulty in finding time to purchase the necessary equipment

D The space in the site office to store the necessary equipment

E The cost of first-aid equipment

**17.23** Which of the following tasks would you expect the appointed person for first aid to carry out?

[A] They should provide most of the care normally carried out by a first aider

[B] They should provide all of the care normally carried out by a first aider

[C] They should contact the emergency services when required

[D] They should only apply splints to broken bones

**17.24** What does the proactive monitoring of health and safety procedures involve?

[A] Ensuring that staff always do the work that they have been instructed to do safely

[B] Deciding how to prevent accidents similar to those that have already occurred

[C] Looking at the work to be done, what could go wrong and how it could be done safely

[D] Checking that all staff read and understand all health and safety notices

**17.25** Why may a young person be more at risk of having an accident?

[A] Legislation does not apply to anyone under 18 years of age

[B] They are usually left to work alone to gain experience

[C] They have less experience and may not recognise danger or understand fully what could go wrong

[D] They are less likely to wear personal protective equipment (PPE)

**17.26** How should cylinders containing liquefied petroleum gas (LPG) be stored on site?

[A] In a locked cellar with clear warning signs

[B] In a locked cage at least 3 m from any oxygen cylinders

[C] In a secure storage container at the back of the site

[D] Covered by a tarpaulin to shield the compressed cylinder from sunlight

**17.27** Where should liquefied petroleum gas (LPG) cylinders supplying an appliance in a site cabin be positioned?

[A] Inside the cabin in a locked cupboard

[B] Under the cabin

[C] Inside the cabin next to the appliance

[D] Outside the cabin

**17.28** What should be used to protect passers-by from getting arc eye when electric welding is about to start on your site?

[A] Warning signs

[B] Screens

[C] Personal protective equipment (PPE)

[D] Nothing

**17.29** What should be the capacity of a spillage bund around a fuel storage tank, in addition to the volume of the tank?

A  10% (110% of the total content)

B  30% (130% of the total content)

C  50% (150% of the total content)

D  75% (175% of the total content)

**17.30** If there is a fatal accident or a reportable dangerous occurrence on site, when **must** the Health and Safety Executive (HSE) be informed?

A  Immediately

B  Within five days

C  Within seven days

D  Within ten days

**17.31** What **must** happen if a prohibition notice is issued by an inspector of the Health and Safety Executive (HSE) or Local Authority?

A  Work can continue, as long as a risk assessment is carried out

B  The work that is subject to the notice must stop

C  The work can continue if extra safety precautions are taken

D  The work in hand can be completed, but no new works started

**17.32** Who should you inform if someone tells you that they have work-related hand-arm vibration syndrome (HAVS)?

A  The Health and Safety Executive (HSE)

B  The local Health Authority

C  The person's doctor

D  The nearest hospital

**17.33** When does an employer have to prepare a written health and safety policy?

A  If they employ five people or more

B  If they employ three people or more

C  If they employ a safety officer

D  If the work is going to last more than 30 days

**17.34** The significant findings of risk assessments **must** be recorded when how many people are employed?

A  Three or more

B  Five or more

C  Six or more

D  Seven or more

17

**17.35** Before allowing a lifting operation to be carried out, where should you ensure that the sequence of operations is recorded to enable a safe lift?

A    In the crane hire contract

B    In an approved lifting plan or method statement

C    In a lifting operation toolbox talk

D    In a risk assessment

**17.36** What does the term lower exposure action value (80 decibels (dBA)) mean, when referring to noise?

A    The average background noise level

B    The noise level at which the worker can request hearing protection

C    The level of noise which must not be exceeded on the site boundary

D    The noise level at which the worker must wear hearing protection

**17.37** At what decibel (dBA) level does it become mandatory for an employer to establish hearing protection zones?

A    80 decibels (dBA)

B    85 decibels (dBA)

C    90 decibels (dBA)

D    95 decibels (dBA)

**17.38** At what minimum noise level must you provide hearing protection to workers if they ask for it?

A    80 decibels (dBA)

B    85 decibels (dBA)

C    87 decibels (dBA)

D    90 decibels (dBA)

**17.39** What is the significance of a weekly or daily personal noise exposure of 87 decibels (dBA)?

A    It is the lower action value and no action is necessary

B    It is the upper action value and hearing protection must be issued

C    It is the peak sound pressure and all work must stop

D    It is the exposure limit value and must not be exceeded

**17.40** In considering what measures to take to protect workers against risks to their health and safety, when should personal protective equipment (PPE) be considered?

A    First, because it is an effective way to protect people

B    First, as the only practical measure

C    Never, as using PPE is bad practice

D    Only when the risks cannot be eliminated by other means

**17.41** In deciding what control measures to take, following a risk assessment that has revealed a risk, what measure should you always consider first?

A  Make sure personal protective equipment (PPE) is available

B  Adapt the work to the individual

C  Give priority to measures that protect the whole workforce

D  Avoid the risk altogether if possible

**17.42** Why is it important that hazards are identified?

A  They have the potential to cause injury or harm

B  They must all be eliminated before work can start

C  They must all be notified to the Health and Safety Executive (HSE)

D  So that toolbox talks can be given on the hazards

**17.43** In the context of a risk assessment, what does the term risk mean?

A  Anything that could cause harm to you or another person

B  Any unsafe act or condition which could cause loss, injury or damage

C  The likelihood that you, or someone else, could be harmed, and how serious any harm could be

D  Any work activity that can be described as hazardous or dangerous

**17.44** Which of the following should be the first consideration if you need to use a hazardous substance?

A  What instruction, training and supervision to provide

B  What health surveillance arrangements will be needed

C  How to minimise risk and control exposure

D  How to monitor the exposure of workers in the workplace

**17.45** What is the purpose of using a permit to work system?

A  To ensure that the job is carried out quickly

B  To ensure that the job is carried out by the easiest method

C  To enable tools and equipment to be properly checked before work starts

D  To establish a controlled, safe system of work

**17.46** If a scaffold is not complete, which of the following actions should be taken by the supervisor?

A  Make sure that the scaffolders complete the scaffold

B  Tell operatives not to use the scaffold

C  Display a warning notice and tell operatives to use the scaffold with care

D  Prevent access to the scaffold and add warning signage

17

**17.47** Following a scaffold inspection under the Work at Height Regulations, how soon **must** a report be given to the person on whose behalf the inspection was made?

A Within two hours

B Within six hours

C Within 12 hours

D Within 24 hours

**17.48** On a scaffold, what is the largest allowable size of an unprotected gap between any guard-rail, toe-board, barrier or other similar means of protection?

A 400 mm

B 470 mm

C 500 mm

D 600 mm

**17.49** What is the **maximum** unprotected gap allowed between any guard-rail, toe-board, barrier or other similar means of protection on a scaffold?

400 mm    470 mm    500 mm    600 mm

**17.50** What is the **minimum** height of the main guard-rail on a scaffold?

875 mm    910 mm    950 mm    1000 mm

**17.51** What is the best way for a supervisor or manager to make sure that the operatives doing a job have fully understood a method statement?

A Put the method statement in a labelled ring-binder in the office

B Explain the method statement to those doing the job and test their understanding

C Make sure that those doing the job have read the method statement

D Display the method statement on a noticeboard in the office

**17.52** Where must the number of people who may be carried in a passenger hoist on site be displayed?

A On a legible notice in the site welfare area

B On a legible notice within the cage of the hoist

C On a legible notice displayed during the site induction

D On a legible notice handed to the hoist operator

**17.53** From a safety point of view, which of the following should be considered first when deciding on the number and location of access and egress points on a site?

A Off-road parking for cars and vans

B Access for the emergency services

C Access for heavy vehicles

D Site security

**17.54** How should access be controlled if people are working in a riser shaft?

A By a site security operative

B By those who are working in it

C By the main contractor

D By a permit to work system

**17.55** What is your least reliable source of information when assessing the level of vibration from a powered hand tool?

A In-use vibration measurement of the tool

B Vibration figures taken from the tool manufacturer's handbook

C Your own judgement based upon observation or experience

D Vibration data from the Health and Safety Executive's (HSE) master list

**17.56** What regulation contains details of the welfare facilities that must be provided on a construction site?

A The Control of Substances Hazardous to Health (COSHH)

B The Construction (Design and Management) Regulations

C The Management of Health and Safety at Work Regulations

D The Workplace (Health, Safety and Welfare) Regulations

17

**17.57** What is regarded as the last resort in the hierarchy of control for operatives' safety when working at height?

A. Safety harness

B. Mobile elevating work platform (MEWP)

C. Mobile access tower

D. Access tower scaffold

**17.58** Which of the following is a fall-arrest system?

A. Guard-rail and toe-board

B. Mobile access tower

C. Mobile elevating work platform (MEWP)

D. Safety harness and lanyard

**17.59** Under the requirements of the Work at Height Regulations, what must the minimum width of a working platform be?

A. Suitable and sufficient for the job in hand

B. Two scaffold boards wide

C. Three scaffold boards wide

D. Four scaffold boards wide

**17.60** When do the Work at Height Regulations require a working platform to be inspected by a competent person?

A. After it has been erected and then at monthly intervals

B. After it has been erected and then at intervals not exceeding 10 days

C. Only after it has been erected

D. After it has been erected and then at intervals not exceeding seven days

**17.61** What is the advantage of using safety nets rather than a harness and fall-arrest lanyard?

A. Safety nets do not need inspecting

B. Workers' lanyards can get entangled with other workers' lanyards

C. Safety nets provide collective fall protection

D. Safety nets can be rigged by anyone

**17.62** What is the maximum vertical height that a fixed ladder can be climbed, before an intermediate landing place is required?

| 7.5 m | 8 m | 8.5 m | 9 m |

**17.63** What should you do if you notice that operatives working above a safety net are dropping off-cuts of material and other debris into the net?

A Nothing, as the debris is all collecting in one place

B Ensure that the net is cleared of debris weekly

C Have the net cleared and inspected, then ensure it is not allowed to happen again

D Ensure that the net is cleared of debris daily

**17.65** When putting people to work above public areas, what should be your first consideration?

A To minimise the number of people below at any one time

B To prevent complaints from the public

C To let the public know what you are doing

D To prevent anything falling onto people below

**17.64** What should be included in a method statement for working at height? Give three answers.

You will be asked to 'drag and drop' your answers

A How falls are to be prevented

B Who will supervise the job on site

C How much insurance cover will be required

D The cost of the job and the time it will take

E The sequence of operations and the equipment to be used

**17.66** Ideally, where should a safety net be rigged?

A Immediately below where you are working

B 2 m below where you are working

C 6 m below where you are working

D At any height below the working position

**17.67** What **must** edge protection be designed to do?

A Allow persons to work on both sides of it

B Secure tools and materials close to the edge

C Warn people where the edge of the roof is

D Prevent people and materials from falling

**17.68** When should guard-rails be fitted to a working platform?

A If it is possible to fall 2 m

B At any height if a fall could result in an injury

C If it is possible to fall 3 m

D Only if materials are being stored on the working platform

**17.69** The Beaufort Scale is important when working at height externally. What does it measure?

A Air temperature

B The load-bearing capacity of a flat roof

C Wind speed

D The load-bearing capacity of a scaffold

**17.70** A design feature of some airbags used for fall arrest is a controlled leak rate. If you are using these, what **must** you ensure about the inflation pump?

A It must be electrically powered

B It must be switched off from time to time to avoid over-inflation

C It must run all the time while work is carried out at height

D It must be switched off when the airbags are full

**17.71** Why is it dangerous to use inflatable airbags for fall arrest if they are too big for the area to be protected?

A They will exert a sideways pressure on anything that is containing them

B The pressure in the bags will cause them to burst

C The inflation pump will become overloaded

D They will not fully inflate

17.72 What is the main danger of leaving someone who has fallen suspended in a harness for too long?

A  The anchorage point may fail

B  They may try to climb back up the structure and fall again

C  They may suffer loss of consciousness and further injury

D  It is a distraction for other workers

# 18  Demolition

**18.01** If asbestos is present, what should happen before demolition or refurbishment takes place?

- A Advise workers that asbestos is present, then continue with the demolition
- B Remove all asbestos as far as is reasonably practicable
- C Advise the Health and Safety Executive (HSE) that asbestos is present, then continue with the demolition
- D Inspect the condition of the asbestos materials

**18.02** What kind of survey is required to identify asbestos prior to demolition?

- A Type 3 survey
- B Management survey
- C Demolition survey
- D Type 2 survey

**18.03** Who **must** be the **first** person a demolition contractor appoints before undertaking demolition operations?

- A A competent person to supervise the work
- B A sub-contractor to strip out the buildings
- C A safety officer to check on health and safety compliance
- D A quantity surveyor to price the extras

**18.04** If there are any doubts about a building's stability, who should a demolition contractor consult?

- A Another demolition contractor
- B A structural engineer
- C A Health and Safety Executive (HSE) factory inspector
- D The company safety adviser

**18.05** Which piece of equipment could a 17-year-old trainee demolition operative use unsupervised?

- A Excavator 360°
- B Dump truck
- C Wheelbarrow
- D Rough terrain forklift

**18.06** When would it be unsafe to operate a scissor lift?

- A If the controls on the platform are used
- B If the ground is soft and sloping
- C If weather protection is not fitted
- D If the machine only has half a tank of fuel

**18.07** On site, what is the minimum distance that oxygen should be stored away from propane, butane or any other gas?

1 m    2 m    3 m    4 m

---

**18.08** What type of fire extinguisher should not be provided where petrol or diesel is being stored?

A. Foam

B. Water

C. Dry powder

D. Carbon dioxide

**18.09** Where should liquefied petroleum gas (LPG) cylinders be located when being used for heating or cooking in site cabins?

A. Under the kitchen work surface

B. Inside but near the door for ventilation

C. In a nearby storage container

D. Securely outside the cabin

**18.10** What is most likely to be caused by continual use of hand-held breakers or drills?

A. Dermatitis

B. Weil's disease (leptospirosis)

C. Vibration white finger

D. Skin cancer

**18.11** What is the most common source of high levels of lead in the blood of operatives during demolition work of an old building?

A. Stripping lead sheeting

B. Cold cutting lead-covered cable

C. Cold cutting fuel tanks

D. Hot cutting coated steel

18

# SPECIALIST

**18.12** Which of the following items of personal protective equipment (PPE) provides the **lowest** level of protection when working in dusty conditions?

- [A] FFP1-rated dust mask
- [B] Positive pressure-powered respirator
- [C] FFP3-rated half mask respirator
- [D] Self-contained breathing apparatus

**18.13** Which of the following would be suitable to use when cutting coated steelwork?

- [A] A disposable dust mask
- [B] A positive pressure-powered respirator
- [C] A high-efficiency dust respirator
- [D] A nuisance dust mask

**18.14** What should you do while reversing mobile plant if you lose sight of the vehicle marshaller who is directing you?

- [A] Carry on reversing slowly
- [B] Stop the vehicle
- [C] Adjust your wing mirror
- [D] Sound the horn and move forward

**18.15** What should you do when leaving mobile plant unattended?

- [A] Leave the engine running, if safe to do so
- [B] Park it in a safe place, remove the keys and lock it
- [C] Put the parking brake on and tell people not to use it
- [D] Put a sign saying 'no unauthorised access' on it

**18.16** Which statement is **true** with regard to using machines?

- [A] Guards can be removed to make work easier
- [B] It's OK to wear rings and other jewellery as long as you take care
- [C] You can carefully remove waste material while the machine is in motion
- [D] Never use a machine unless you have been trained and given permission to use it

**18.17** Which of the following is **not** generally part of a plant operator's daily pre-use check?

- [A] Emergency systems
- [B] Engine oil level
- [C] Hydraulic fluid level
- [D] Brake pad wear

**18.18** Which of these statements is true in relation to an operator of a scissor lift?

A They must be trained and authorised in the use of the equipment

B They must only use the ground level controls

C They must be in charge of the work team

D They must ensure that only one person is on the platform at any time

**18.19** On demolition sites, what must the drivers of plant have, for their own and others' safety?

A Adequate visibility from the driving position

B A temperature controlled cab

C Wet weather gear for when it's raining

D A supervisor in the cab with them

**18.20** When must head and tail lights be used on mobile plant?

A If the plant is using the same traffic route as private cars

B When the plant is operating in conditions of poor visibility

C When the plant is operated by a trainee

D Only if the plant is crossing pedestrian routes

**18.21** What safety feature is provided by FOPS on mobile plant?

A The speed is limited when tracking over hard surfaces

B The machine stops automatically if the operator lets go of the controls

C The operator is protected from falling objects

D The reach is limited when working near to live overhead cables

**18.22** Where is the only place you will not find information about the daily checks required for mobile plant?

A On stickers attached to the machine

B In the manufacturer's handbook

C In the supplier's information

D On the health and safety law poster

**18.23** What should you do if you discover underground services not previously identified?

A Fill in the hole immediately

B Stop work until the situation has been resolved

C Cut the pipe or cable to see if it's live

D Get the machine driver to dig it out

18

**18.24** What action should you take if you discover unlabelled drums or containers on site?

- [A] Put them in the nearest waste skip
- [B] Ignore them, as they will get flattened during the demolition
- [C] Stop work until they have been safely dealt with
- [D] Open them and smell the contents to see if they are flammable

**18.25** If the plant you are driving has defective brakes, what action should you take?

- [A] Reduce your speed
- [B] Report it and carry on working
- [C] Report it and isolate the machine
- [D] Use the handbrake instead of the foot brake

**18.26** What action should be taken if a wire rope sling is defective?

- [A] Do not use it and make sure that no-one else can
- [B] Only use it for up to half of its safe working load
- [C] Report the defect at the end of the day
- [D] Only use it for small lifts under 1 tonne

**18.27** With regard to the safe method of working, what is the **most** important subject of induction training for demolition operatives?

- [A] Working hours on the site
- [B] Explanation of the method statement
- [C] Location of welfare facilities
- [D] COSHH assessments

**18.28** Which **two** of the following documents refer to the specific hazards associated with demolition work in confined spaces?

- [A] Safety policy
- [B] Permit to work
- [C] Risk assessment
- [D] Scaffolding permit
- [E] Hot-work permit

**18.29** When asbestos material is suspected in buildings to be demolished, what is the **first** priority?

- [A] Ensure a competent person carries out an asbestos survey
- [B] Notify the Health and Safety Executive (HSE) of the possible presence of asbestos
- [C] Remove and dispose of the asbestos
- [D] Employ a licensed asbestos remover

**18.30** What is the safest method of demolishing brick or internal walls by hand?

A Undercut the wall at ground level

B Work across in even courses from the ceiling down

C Work from the doorway at full height

D Cut down at corners and collapse in sections

**18.31** Who should be consulted before demolition is carried out near to overhead cables?

A The Health and Safety Executive (HSE)

B The fire service

C The electricity supply company

D The land owner

**18.32** When demolishing a building in controlled sections, what is the most important consideration for the remaining structure?

A The soft strip is completed

B All non-ferrous metals are removed

C It remains stable

D Trespassers cannot get in at night

**18.33** Where would you find out the method for controlling identified hazards on a demolition project?

A The demolition toolbox

B The health and safety file

C The pre-tender health and safety plan

D The construction phase plan

**18.34** When hinge-cutting a steel building or structure for a controlled collapse, which of the following should be the last cuts?

A Front leading row top cuts

B Front leading row bottom cuts

C Back row top cuts

D Back row bottom cuts

**18.35** What should be obtained before carrying out the demolition cutting of fuel tanks?

A A gas free certificate

B An environmental certificate

C A risk assessment

D A COSHH assessment

18

**18.36** What do the letters SWL stand for?

A   Satisfactory working limit

B   Safe working level

C   Satisfactory weight limit

D   Safe working load

**18.37** Which of the following is true in relation to the safe working load of a piece of equipment?

A   It must never be exceeded

B   It is a guide figure that may be exceeded slightly

C   It may be exceeded by 10% only

D   It gives half the maximum weight to be lifted

**18.38** What should be clearly marked on all lifting gear?

A   Date of manufacture

B   Name of maker

C   Date next test is due

D   Safe working load

**18.39** How often should lifting accessories be thoroughly examined?

A   At least every three months

B   At least every six months

C   At least every 14 months

D   At least every 18 months

**18.40** What is the correct way to climb off a machine?

A   Jump down from the seated position

B   Climb down, facing forward

C   Climb down, facing the machine

D   Use a ladder

**18.41** When is it acceptable to carry passengers on a machine?

A   When the employer gives permission

B   When they are carried in the skip

C   When the machine is fitted with a purpose-made passenger seat

D   When the maximum speed is no greater than 10 mph

18

# 19 Highway works

**19.01** Why is it not safe to use diesel to prevent asphalt sticking to the bed of lorries?

- A It will create a slipping hazard
- B It will corrode the bed of the lorry
- C It will create an environmental hazard
- D It will react with the asphalt, creating explosive fumes

**19.02** You are moving or laying slabs, paving blocks or kerbs on site. Which two of the following methods would be classed as manual handling?

- A Using a scissor lifter attachment on an excavator
- B Using a trolley or sack barrow
- C Using a suction lifter attached to an excavator
- D Using a two man scissor lifter
- E Using a crane with fork attachment

**19.03** What are two effects of under-inflated tyres on the operation of a machine?

- A It decreases the operating speed of the engine
- B It can make the machine unstable
- C It causes increased tyre wear
- D It causes decreased tyre wear
- E It increases the operating speed of the engine

**19.04** If you are driving any plant that may overturn, when must a seat belt be worn?

- A When travelling over rough ground
- B When the vehicle is loaded
- C When you are carrying passengers
- D At all times

**19.05** If a safety cut-out on a machine does not operate, what should you do?

- A Keep quiet in case you get the blame
- B Report it at the end of the shift
- C Try and fix it or repair it yourself
- D Stop and report it immediately to your supervisor

**19.06** When can you carry authorised passengers in vehicles?

- A Only if your supervisor gives you permission
- B Only if a suitable secure seat is provided for each of them
- C Only when off the public highway
- D Only if you have a full driving licence

19

**19.07** When it is necessary to tip into an excavation, what is the preferred method of preventing the vehicle getting too close to the edge?

A Signage

B Driver's experience

C Signaller

D Stop blocks

**19.08** Which of these is the safest method of operating a machine under electrical power lines when it has an extending jib or boom?

A Erect signs to warn drivers while they are operating the machines

B Adapt the machine to limit the extension of the jib or boom

C Place a warning notice on the machine

D Let down the tyres on the machine to increase the clearance

**19.09** When using lifting equipment, such as a cherry picker, lorry loader or excavator, how should you follow the indicated safe working load?

A It must never be exceeded

B It is a guide figure that may be exceeded slightly

C It may be exceeded by 10% only

D It gives half the maximum weight to be lifted

**19.10** Which checks should the operator of a mobile elevating work platform (MEWP) (for example, a cherry picker) carry out before using it?

A That a seat belt is provided for the operator

B That a roll-over cage is fitted

C That the hydraulic system is drained

D That emergency systems operate correctly

**19.11** When should you switch on the amber flashing beacon fitted to your highways vehicle?

A At all times

B When travelling to and from the depot

C When it is being used as a works vehicle

D Only in poor visibility

**19.12** What must you do approximately 200 m before a site works access on a motorway?

A Switch on the vehicle hazard lights

B Switch on the flashing amber beacon

C Switch on the headlights

D Switch on the flashing amber beacon and the appropriate indicator

19

**19.13** How often **must** a competent person thoroughly examine lifting equipment for lifting persons (for example, a cherry picker)?

A Every six months

B Every 12 months

C Every 18 months

D Every 24 months

**19.14** What **must** you do before towing a trailer fitted with independent brakes?

A Fit a safety chain

B Fit a cable that applies the trailer's brakes if the tow hitch fails

C Use a rope to secure the trailer to the tow hitch

D Make sure it is possible to drive at a maximum speed of 25 mph along the route you have chosen

**19.15** What **must** you do when getting off plant and vehicles?

A Use a pair of steps which allow you to climb out of the vehicle

B Use the wheels and tyres for access

C Use the designated access and egress point whilst maintaining three points of contact with the vehicle

D Look before you jump, then jump down facing the vehicle

**19.16** What should you do when leaving plant unattended?

A Leave the amber flashing beacon on

B Apply the brake, switch off the engine and remove the key

C Leave it in a safe place with the engine ticking over

D Park with blocks under the front wheels

**19.17** What **must** you have before towing a compressor on the highway? Give **two** answers.

A Permission from your supervisor

B The correct class of driving licence

C Permission from the police

D Permission from the compressor hire company

E Working lights, and a number plate on the compressor that matches that of the towing vehicle

**19.18** Who is responsible for the security of the load on a vehicle?

A The driver's supervisor

B The police

C The driver

D The driver's company

**19.19** What is the correct way to dismount from the cab or driving seat of plant and vehicles?

A. Use three points of contact facing forwards (away from the vehicle)

B. Jump down facing forwards (away from the vehicle)

C. Use three points of contact facing backwards (towards the vehicle)

D. Jump down well clear of the vehicle

**19.20** A single vehicle is being used to carry out mobile highway works during the day. What sign or symbol **must** be clearly displayed on or at the rear of the vehicle?

A. A road narrows sign (left or right)

B. A specific task warning sign (for example, gully cleaning)

C. A keep left or right arrow

D. A roadworks ahead sign

**19.21** Why is it necessary to wear hi-vis clothing when working on roads?

A. So road users and plant operators can see you

B. So your supervisor can see you

C. To protect clothes worn underneath from damage

D. Because it will keep you warm

**19.22** You are working on a dual carriageway with a 60 mph speed limit although you are not in the working space. What is the **minimum** standard of hi-vis clothing that you **must** wear?

A. Reflective waistcoat

B. Reflective long-sleeved jacket

C. Reflective sash

D. Reflective hard hat

**19.23** When kerbing works are being carried out, how should kerbs be taken off the vehicle?

A. By lifting them off manually using the correct technique

B. By pushing them off the back

C. By using mechanical means, such as a machine fitted with a grab

D. By asking your workmate to give you a hand

**19.24** In which **two** places would you find information about the distances for setting out highways signs in advance of the works under different road conditions?

A. In the Traffic Signs Manual (Chapter 8)

B. In the Pink Book

C. On the back of the sign

D. In the Works Order

E. In the new Code of Practice (Red Book)

**19.25** What should you do with materials that have to be kept on site overnight?

A Don't stack them above 2 m high

B Stack them safely and in a secure area

C Put pins and bunting around them

D Only stack them on the grass verge

**19.26** What is the purpose of an on-site risk assessment?

A To ensure there is no risk of traffic build-up due to the works in progress

B To identify hazards and risks specific to the site in order to ensure a safe system of work

C To ensure that the work can be carried out in reasonable safety

D To protect the employer from prosecution

**19.27** Where would you find information about working on a dual carriageway with a speed limit above 40 mph?

A In the welfare cabin

B In the Blue Book

C In the Traffic Signs Manual (Chapter 8)

D In the Pink Book

**19.28** What is not usually something to consider while undertaking a site-specific risk assessment before starting highway works?

A The cost of the sub-contractor who will be carrying out the work

B The amount and type of traffic

C The effect of different weather conditions

D The type and size of the road

**19.29** What should you do if you are the driver of a vehicle and lose sight of the vehicle marshaller while reversing your vehicle?

A Continue reversing slowly

B Continue reversing, as long as your vehicle is equipped with a klaxon and flashing lights

C Stop and locate the vehicle marshaller

D Find someone else to watch you reverse safely

**19.30** When providing portable traffic signals on roads used by cyclists, what action should you take?

A Locate the signals at bends in the road

B Allow more time for slow-moving traffic by increasing the all-red phase of the signals

C Operate the signals manually

D Use stop-and-go boards only

19

**19.31** What is the **main** reason why temporary highways signing needs to be removed when works are completed?

A It gets traffic flowing

B It is a legal requirement

C To allow the road to be opened fully

D To reuse signs on new jobs

**19.32** Which **two** site conditions must be met before traffic management can be reduced to the **minimum** requirements?

A Traffic is heavy

B Visibility is good

C There are double yellow lines

D There is a footpath

E It is a period of low risk

**19.33** What traffic management is required when carrying out a maintenance job on a motorway?

A The same as would be required on a single carriageway

B A flashing beacon and a keep left or right sign

C A scheme installed by a registered traffic management contractor

D Ten 1 m high cones and a 1 m high 'men working' sign

**19.34** What is the **minimum** traffic management required when carrying out a short-term minor maintenance job in a quiet, low-speed side road?

A A flashing amber beacon and a keep left or right arrow

B The same as required for a road excavation

C Five cones and a blue arrow

D Temporary traffic lights

**19.35** What is the **maximum** distance between the 'roadworks ahead' signs for work activities that move along the carriageway, such as sweeping, verge mowing and road lining?

A Quarter of a mile

B Half a mile

C One mile

D Two miles

**19.36** What action is required when a highways vehicle fitted with a direction arrow is travelling from site to site?

A Point the direction arrow up

B Travel slowly from site to site

C Point the direction arrow down

D Cover or remove the direction arrow

**19.37** How **must** signs on footways be located?

A So that they block the footway

B So that they can be read by site personnel

C So that they do not create a hazard for pedestrians

D So that they can be easily removed

**19.38** What should you do if drivers approaching highway works **cannot** see the advance signs clearly because of poor visibility or obstructions caused by road features?

A Place additional signs in advance of the works

B Extend the safety zones

C Extend the sideways clearance

D Lengthen the lead-in taper

**19.39** How should you protect a portable traffic-light cable that crosses a road?

A It should be secured firmly to the road surface

B A cable crossing protector must be used with ramp warning signs

C It can be unprotected if it is less than 10 mm in diameter

D It should be placed in a slot cut into the road surface

**19.40** What action is required where passing traffic may block the view of highways signs?

A Signs must be larger

B Signs must be duplicated on both sides of the road

C Signs must be placed higher

D Additional signs must be placed in advance of the works

**19.41** Which is **not** an approved means of controlling traffic at roadworks?

A Priority signs

B Police supervision

C Hand signals by operatives

D A give-and-take system

**19.42** What action is required if a vehicle detector on temporary traffic lights becomes defective?

A Control traffic at the defective end using hand signals

B Operate on all-red and call the service engineer

C Operate on fixed time or manual and call the service engineer

D Switch the lights off until the supervisor arrives on site

19

**19.43** How should portable traffic signals be assembled and placed?

A As speedily as possible

B In an organised manner, to a specified sequence

C During the night

D As work starts each morning

**19.44** What action is required where it is not possible to maintain the correct safety zone?

A Barrier off the working space

B Place additional advance signing

C Use extra cones on the lead-in taper

D Stop work and consult your supervisor

**19.45** In which of the following circumstances can someone enter the safety zone?

A To store unused plant

B To maintain cones and signs

C To park site vehicles

D To store materials

**19.46** What action should you take if a vehicle, driven by a member of the public, enters the coned off area on a dual carriageway?

A Remove a cone and direct the driver back on to the live carriageway

B Ignore them

C Shout and wave them off site

D Help them to leave the site safely using the nearest designated exit

**19.47** When working after dark, is mobile plant exempt from the requirement to show lights?

A Yes, always

B Yes, if authorised by the supervisor

C Only if they are not fitted to the machine as standard

D Not in any circumstances

**19.48** What is the purpose of the safety zone?

A To indicate the works area

B To protect you from the traffic and the traffic from you

C To allow extra working space in an emergency

D To give a safe route around the working area

**19.49** What should be used to protect the public from a shallow excavation in a public footway?

A  Pins and bunting

B  Nothing

C  Cones

D  Barriers with tapping rails

**19.50** When should installed highways signs and guarding equipment be inspected?

A  Immediately after it has been used

B  No more frequently than once a week

C  Every hour, except when the site is unattended

D  Regularly, and at least once every day, including when the site is unattended

**19.51** How must highway signs, lights and guarding equipment be properly secured?

A  By built-in weights where possible

B  By roping them to concrete blocks or kerb stones

C  By pushing them securely into the soft verge

D  By iron weights suspended from the frame by chains or other strong material

**19.52** In which of the following circumstances would it not be safe to use a cherry picker for working at height?

A  When a roll-over cage is not fitted

B  When the ground is uneven and sloping

C  When weather protection is not fitted

D  When the operator is clipped to an anchorage point in the basket

19

## 20 Specialist work at height

**20.01** If you need to store materials on a roof, what **three** things **must** you do?

- [A] Check the load bearing capability of the roof to avoid damage to the structure
- [B] Stack materials no more than 1.2 m above the guard-rail height
- [C] Ensure there is safe access and clear working areas around the materials for everyone working on the roof
- [D] Use a gin wheel and rope tied to a temporary tripod at the roof edge for raising and lowering the materials
- [E] Store the materials in a way that prevents them from falling off, or being blown off, the roof

**20.02** What should you do if a safety lanyard has damaged stitching?

- [A] Use the lanyard if the damaged stitching is less than 5 cm long
- [B] Get a replacement lanyard before starting work
- [C] Not use the damaged lanyard and work without one
- [D] Use the lanyard if the damaged stitching is less than 15 cm long

**20.03** What is the **main** danger of leaving someone who has fallen suspended in a harness for too long?

- [A] The anchorage point may fail
- [B] They may try to climb back up the structure and fall again
- [C] They may suffer loss of consciousness and further injury
- [D] It is a distraction for other workers

**20.04** If using inflatable airbags as a means of fall arrest, what **must** you ensure with regard to the inflation pump?

- [A] It must be electrically powered
- [B] It must be switched off from time to time to avoid over-inflation
- [C] It must run all the time while work is carried out at height
- [D] It must be switched off when the airbags are full

**20.05** Why is it dangerous to use inflatable airbags that are too big for the area to be protected?

- [A] They will exert a sideways pressure on anything that is containing them
- [B] The pressure in the bags will cause them to burst
- [C] The inflation pump will become overloaded
- [D] They will not fully inflate

**20.06** When is it **most** appropriate to use a safety harness and fall-arrest lanyard for working at height?

- [A] Only when the roof has a steep pitch
- [B] Only when crossing a flat roof with clear roof lights
- [C] Only when all other options for fall prevention have been ruled out
- [D] Only when materials are stored at height

**20.07** When trying to clip your lanyard to an anchor point you find the locking device does **not** work. What should you do?

- A Carry on working and report it later
- B Tie the lanyard in a knot round the anchor
- C Stop work and report it to your supervisor
- D Carry on working without it

**20.08** What is the **main** reason for using a safety net or other soft-landing system rather than a personal fall-arrest system?

- A Soft-landing systems are cheaper to use and do not need inspecting
- B It is always easier to rescue workers who fall into a soft-landing system
- C Specialist training is not required to install soft-landing systems
- D Soft-landing systems are collective fall arrest measures

**20.09** What is edge protection designed to do?

- A Make access to the roof easier
- B Secure tools and materials close to the edge
- C Stop rainwater running off the roof onto workers below
- D Prevent people and materials from falling

**20.10** What is the **maximum** permitted gap between the guard-rails on a working platform?

350 mm      410 mm      470 mm      520 mm

20

**20.11** When should guard-rails be fitted to a working platform?

A  If it is possible to fall 2 m

B  At any height if a fall could result in an injury

C  If it is possible to fall 3 m

D  Only if materials are being stored on the working platform

**20.12** The Beaufort Scale is important when working at height externally. What does it measure?

A  Air temperature

B  The load-bearing capacity of a flat roof

C  Wind speed

D  The load-bearing capacity of a scaffold

**20.13** Before starting work at height, the weather forecast says the wind will increase to Force 7. What is the **best** description of the wind conditions?

A  A moderate breeze that can raise light objects, such as dust and leaves

B  A near gale that will make it difficult to move about and handle materials

C  A gentle breeze that you can feel on your face

D  Hurricane winds that will uproot trees and cause structural damage

**20.14** If you have to lean over an exposed edge while working at height, how should you wear your safety helmet?

A  Tilted back on your head so that it doesn't fall off

B  Take your helmet off while leaning over then put it on again afterwards

C  Wear the helmet as usual but use the chinstrap

D  Wear the helmet back to front whilst leaning over

**20.15** Before climbing a ladder you notice that it has a rung missing near the top. What should you do?

A  Do not use the ladder, and immediately report the defect

B  Use the ladder but take care when stepping over the position of the missing rung

C  Turn the ladder over so that the missing rung is near the bottom and then use it

D  See if you can find a piece of wood to replace the rung

**20.16** How far should a ladder extend above the stepping-off point if there is no alternative, firm handhold?

A  One rung

B  Two rungs

C  One metre

D  Half a metre

20

**20.17** When using portable or pole ladders for access, what is the maximum vertical distance between landings?

A There is no maximum

B 4 m

C 9 m

D 30 m

**20.18** You need to use a ladder to access a roof but the only place to rest the ladder is on a run of plastic gutter. What two things should you consider doing?

A Resting the ladder on a gutter support bracket

B Resting the ladder against the gutter, climbing it and quickly tying it off

C Finding another way to access the roof

D Using a proprietary stand-off device that allows the ladder to rest against the wall

E Positioning the ladder at a shallow angle so that it rests below the gutter

**20.19** What should someone working from a cherry picker attach their lanyard to?

A A strong part of the structure that they are working on

B A secure anchorage point inside the platform

C A secure point on the boom of the machine

D A scaffold guard-rail

**20.20** You are working at height from a cherry picker when the weather becomes very windy. What should your first consideration be?

A Tie all lightweight objects to the handrails of the basket

B Clip your lanyard to the structure that you are working on

C Tie the cherry picker basket to the structure you are working on

D Decide whether the machine will remain stable

**20.21** If you are on a cherry picker but it does not quite reach where you need to work, what should you do?

A Use a stepladder balanced on the machine platform

B Extend the machine fully and stand on the guard-rails

C Abandon the machine and use a long extending ladder

D Do not carry out the job until you have an alternative means of access

**20.22** If you are working at height and operating a mobile elevating work platform (MEWP), when is it acceptable for someone to use the ground-level controls?

A If the person on the ground is trained and you are not

B In an emergency

C If you need to jump off the MEWP to gain access to the work

D If you need both hands free to carry out the job

20

**20.23** When is it acceptable to jump off a mobile elevating work platform (MEWP) on to a high level work platform?

A If the work platform is fitted with edge protection

B If the machine operator stays in the basket

C Not under any circumstances

D If the machine is being operated from the ground-level controls

**20.24** How will you know the maximum weight or number of people that can be lifted safely on a mobile elevating work platform (MEWP)?

A The weight limit is reached when the platform is full

B It will be stated on the health and safety law poster

C You will be told during site induction

D It will be stated on an information plate fixed to the machine

**20.25** When is it safe to use a mobile elevating work platform (MEWP) on soft ground?

A When the ground is dry

B When the machine can stand on scaffold planks laid over the soft ground

C When ground load bearing capacity has been assessed as suitable

D Never

**20.26** If you need to cross a fragile roof, how do you establish if it is fragile?

A Tread gently and listen for cracking

B Stop and seek advice

C Look at the roof surface and make your own assessment

D It does not matter if you walk along a line of bolts

**20.27** If you notice some overhead cables within reach after gaining access to a roof, what should you do?

A Keep away from them while you work but remember that they are there

B Stop work and confirm that it is safe for you to be on the roof

C Make sure that you are using a wooden ladder

D Hang coloured bunting from them to remind you they are there

**20.28** If you are working above a safety net and you notice the net is damaged, what should you do?

A Work somewhere away from the damaged area of net

B Stop work and report it

C Tie the damaged edges together using the net test cords

D See if you can get hold of a harness and lanyard

**20.29** What is the main reason for not allowing debris to gather in safety nets?

A  It will overload the net

B  It looks untidy from below

C  It could injure someone who falls into the net

D  Small pieces of debris may fall through the net

**20.30** What should you do if you are working at height, but the securing cord for a safety net is in your way?

A  Untie the cord, carry out your work and tie it up again

B  Untie the cord, but ask the net riggers to re-tie it when you have finished

C  Tell the net riggers that you are going to untie the cord

D  Leave the cord alone and report the problem

**20.31** Ideally, where should a safety net be rigged?

A  Immediately below where you are working

B  2 m below where you are working

C  6 m below where you are working

D  At any height below the working position

**20.32** Who should install safety nets?

A  A scaffolder

B  Someone who has had experience of working with them before

C  A trained, competent and authorised person

D  A steel or cladding erector

**20.33** When can someone who is not a scaffolder remove parts of a scaffold?

A  Only if the scaffold is not more than two lifts in height

B  As long as a scaffolder refits the parts after the work has finished

C  Never, as only competent scaffolders can remove the parts

D  Only if they think the parts won't weaken the scaffold

**20.34** What should you do if you find that a scaffold tie is in your way when you are working?

A  Ask a scaffolder to remove it

B  Remove it yourself and then replace it when you have finished

C  Remove it yourself but get a scaffolder to replace it when you have finished

D  Report the problem to your supervisor

20

**20.35** Which type of scaffold tie can be removed by someone who is not a scaffolder?

- A  A box tie

- B  A ring tie

- C  A reveal tie

- D  No types of tie

**20.36** What should be included in a safety method statement for working at height? Give three answers.

- A  The cost of the job and the time it will take

- B  The sequence of operations and the equipment to be used

- C  How much insurance cover will be required

- D  How falls are to be prevented

- E  Who will supervise the job on site

**20.37** When it is not possible to avoid working above public areas, what should be your first consideration?

- A  To minimise the number of people below at any one time

- B  To prevent complaints from the public

- C  To let the public know what you are doing

- D  To prevent anything falling onto people below

**20.38** Roof light covers should meet which two of the following requirements?

- A  They must be made from the same material as the roof covering

- B  They must be made from clear material to allow the light through

- C  They must be strong enough to take the weight of any load placed on them

- D  They must be waterproof and windproof

- E  They must be fixed in position to stop them being dislodged

**20.39** If you need to inspect pipework at high level above an asbestos roof, how should you access it?

- A  Use an extension ladder and crawler board to get to the pipework

- B  Use a ladder to get onto the roof and walk the bolt line on the roof sheets

- C  Report the pipework as unsafe

- D  Hire in suitable mobile access equipment

**20.40** You have been asked to erect specialist access frames using anchor bolts. Before you start work what should you not do?

- A  Check the access frames are sound

- B  Assume that the access system is safe to use

- C  Test the anchor bolts

- D  Ensure that your assistant has their harness on

**20.41** Which of these **must** happen before any roof work starts?

- A) A risk assessment must be carried out following a hierarchy of controls
- B) The operatives working on the roof must be trained in the use of safety harnesses
- C) Permits to work must be issued to those allowed to work on the roof
- D) A weather forecast must be obtained

**20.42** When working at height, what is the **safest** way to transfer waste materials to ground level?

- A) Through a waste chute directly into a skip
- B) Asking someone below to keep the area clear of people, then throwing the waste down
- C) Erecting barriers around the area where the waste will land
- D) Bagging up the waste before throwing it down

**20.43** Who should erect and dismantle scaffold towers?

- A) Someone who has the instruction book
- B) Someone who is trained, competent and authorised
- C) Advanced scaffolders
- D) Someone who has worked on them before

**20.44** After gaining access to the platform of a correctly erected mobile access tower, what is the **first** thing you should do?

- A) Check that the tower's brakes are locked on
- B) Check for overhead power lines
- C) Close the access hatch to stop people or equipment from falling
- D) Check that the tower does not rock or wobble

**20.45** What **must** you do before a mobile access tower is moved?

- A) Clear the platform of people and equipment
- B) Get a permit to work
- C) Get approval from the principal contractor
- D) Make arrangements with the forklift truck driver

**20.46** An outdoor tower scaffold has stood overnight in high winds and heavy rain. What should you ensure before the scaffold is used?

- A) That the brakes still work
- B) That the scaffold is tied to the adjacent structure
- C) That the scaffold is inspected by a competent person
- D) That the platform hatch still works correctly

20

SPECIALIST

**20.47** What is the recommended maximum height for a free-standing mobile tower when it is used indoors?

A  There is no restriction

B  Three lifts

C  The height recommended by the manufacturer

D  Three times the longest base dimension

# 21 Lifts and escalators

**21.01** Who is allowed to safely release trapped passengers?

- A The site manager
- B Only a trained and authorised person
- C Anyone
- D Only the emergency services

**21.02** How should you connect a car light supply to a 240 volt supply (240 volt fused spur)?

- A Connect it with the power on
- B Switch off the spur and then connect it
- C Switch off the spur, remove the fuse and then connect it
- D Isolate and lock off the incoming supply and then connect it

**21.03** If a switch needs to be changed in the pit but the isolator is in the machine room 12 floors above, what should you do?

- A Isolate the power and then lock and tag the isolator
- B Risk assess the situation and change the switch with the power on because it is control voltage
- C Use insulated tools
- D Stand on a rubber mat

**21.04** What is the main cause of injury and absence for workers in the lift and escalator industry?

- A Falls
- B Electrocution
- C Contact with moving parts
- D Manual handling

**21.05** If a counterweight screen is not fitted or has been removed, what should you do before starting work?

- A Carry out a further risk assessment to establish a safe system of work
- B Nothing – just get on with the job as normal
- C Give a toolbox talk on guarding
- D Issue and wear appropriate personal protective equipment (PPE)

**21.06** Which of the following types of fire extinguisher should not be used if there is a fire in a lift or escalator controller?

- A Dry chemical
- B Water
- C Dry powder
- D Carbon dioxide

21

**21.07** What should you do if the lifting accessory you are about to use is defective?

A. Only use it for half its safe working load

B. Only use it for small lifts under 1 tonne

C. Do not use it and inform your supervisor

D. Try to fix it

**21.08** If landing doors are not fitted to a lift on a construction site, what is the minimum height of the barrier that must be fitted instead?

A. 650 mm

B. 740 mm

C. 810 mm

D. 950 mm

**21.09** A set of chain blocks has been delivered to site with an examination report stating that they were examined by a competent person a month ago. The hook is obviously damaged. What action do you take?

A. Use the blocks as the examination report is current

B. Do not use the blocks and inform your supervisor

C. Use the blocks at half the safe working load

D. Use the blocks until replacement equipment arrives

**21.10** When must you not wear rings, bracelets, wrist watches, necklaces and similar items?

A. When working near or on electrical or moving equipment

B. When working on site generally

C. When driving a company vehicle

D. After leaving home for work

**21.11** What is the correct method for disposing of used or contaminated oil?

A. Decant it into a sealed container and place in a skip

B. Dispose of it through a registered waste process

C. Dilute it with water and pour it down a sink

D. Pour it down a roadside drain

**21.12** A large, heavy, balance weight frame is delivered to site on a lorry with no crane and there is no lifting equipment available on site. What should you do?

A. Unload it manually

B. Arrange for it to be re-delivered on a suitable lorry

C. Slide it down planks

D. Tip the load off the lorry

**21.13** A lifting beam at the top of the lift shaft is marked with a safe working load of 800 kg but the brickwork around the beam is cracked and appears to be loose. What should you do?

A  Use the beam as normal

B  Only lift loads not exceeding 400 kg

C  Not use the beam and speak to your supervisor

D  De-rate the beam by 75%

**21.14** What is fitted to prevent injury from an overspeed governor?

A  A rope

B  A restrictor

C  A guard

D  A switch

**21.15** If the escalator or passenger conveyor has an external machine room, which statement applies to its access doors?

A  They should be capable of being locked from both sides and be marked with an appropriate safety sign

B  They should be smoke proof in case of a fire

C  They should be unlocked at all times in case of an emergency

D  They should be capable of being locked on the inside only and be marked with the appropriate safety sign

**21.16** What is the statutory period of examination for lifting equipment that is used to lift people?

A  At least monthly

B  At least every six months

C  At least every 12 months

D  Once every two years

**21.17** What checks do you need to carry out before using lifting equipment?

A  A drop check

B  That it is free from defects and has a current examination certificate

C  That the chains are knotted to the correct length

D  That the lifting tackle states the date of manufacture

**21.18** Following the initial inspection, how often should a scaffold in a lift shaft be inspected by a competent person?

A  At least every day

B  At least every seven days

C  At least every 14 days

D  There is no set period between inspections

21

**21.19** When installing a new rope, what should you do if you notice a damaged section where something heavy has fallen onto the coil?

- A  Fit the rope anyway

- B  Cut out the damaged section

- C  Reject the rope

- D  Add an extra termination

**21.20** What must you do **first**, before entering the pit of an operating lift?

- A  Fit pit props

- B  Verify the pit stop switch

- C  Switch the lift off

- D  Position the access ladder

**21.21** Who should fit a padlock and tag to an electrical lock-out guard?

- A  Anyone authorised to work on the unit

- B  Only the person who fitted the lock-out guard

- C  Only the senior engineer

- D  Only the manufacturer

**21.22** If you arrive on site and find the lift mains isolator switched off, what should you do?

- A  Switch it on and get on with your work

- B  Switch it on and check the safety circuits to see if there is a fault

- C  Contact the person in control of the premises to find out if they had switched it off

- D  Shout down the shaft and, if no-one responds, switch it on and get on with your work

**21.23** Which **two** of the following actions must be carried out by an authorised person working alone?

- A  Registering their presence with the site representative before starting work

- B  Ensuring their timesheet is accurate and countersigned

- C  Establishing suitable arrangements to ensure the monitoring of their wellbeing

- D  Notifying the site manager of the details of their work

- E  Ensuring that the lift pit is free from water and debris

**21.24** Which statement is **true** when using an authorised lifting accessory marked with its safe working load?

- A  Never exceed the safe working load

- B  The safe working load is only for guidance

- C  Halve the safe working load if the equipment is damaged

- D  Double the safe working load if people need to be lifted

21

**21.25** What should be fitted to the main sheave and diverter to prevent injury from rotating equipment?

A  Movement sensors

B  Guards

C  Clutching assemblies

D  Safety notices

**21.26** What are the appropriate types of tools and equipment for working on electrical lift-control equipment?

A  Insulated tools and an insulating mat

B  Non-insulated tools

C  Any tools and an insulating mat

D  No tools are allowed near electrical equipment

**21.27** When installing a partially enclosed or observation lift, what safe system of work can you use to prevent injury to people below?

A  Put up a sign

B  Do not use heavy tools

C  Secure tools to prevent them falling off

D  Only carry out essential work using minimum tools

**21.28** What needs to be checked before any hot work takes place in the lift installation?

A  The weight and size of the welding equipment

B  How long the task will take

C  If a hot-work permit is required

D  If the local fire services need to be notified

**21.29** What must you ensure if the trapdoor or hatch has to be left open while you work in the machine room?

A  That a sign is posted to warn others that you are working there

B  That the distance from the trapdoor or hatch to the floor below does not exceed 2 m

C  That there is sufficient light available for the work

D  That a suitable barrier is put in place around the trapdoor or hatch

**21.30** What must you ensure to prevent unauthorised access to unoccupied machine equipment space?

A  That the access door is locked

B  That a sign is posted to warn trespassers

C  That the power supply is isolated

D  That a person is posted to prevent access

21

**21.31** What should be applied to the main isolator of a traction lift to prevent it starting accidentally?

A | A warning notice

B | A lock-out device

C | A residual current device (RCD)

D | Lower-rated fuses

**21.32** Who is responsible for the keys when a padlock has been applied to a lock-out device?

A | The individual applying the lock

B | The site supervisor

C | The site manager

D | The person nearest the lock-out device

**21.33** If the main contractor wants to use an unfinished lift to move some equipment to an upper floor, what should you do?

A | Help to ensure the load is correctly positioned

B | Tell them to talk to your supervisor

C | Ask for the weight of the equipment

D | Allow them to use the lift but take no responsibility for any accidents

**21.34** What is an essential action before gaining access into the escalator or passenger conveyor?

A | That the mains switch is locked out and tagged

B | That the mains switch is in the On position

C | That all steps are removed

D | That the drive mechanism is lubricated

**21.35** What is secured at the entry and exit points of an escalator or passenger conveyor to prevent people falling into the machine or machine space?

A | Safety barriers

B | Safety notices

C | Escalator machine equipment guards

D | Machine tank covers

**21.36** What **must** you do before moving the steps or pallet band of an escalator or passenger conveyor?

A | Check that there are no sharp edges on the steps

B | Check that there is a clear route of escape

C | Check that no unauthorised people are on the equipment

D | Check that a fire extinguisher is available

**21.37** What is the minimum size of gap between the edge of the work platform and the hoist way wall that is regarded as a fall hazard?

A  250 mm

B  300 mm

C  330 mm

D  450 mm

**21.38** When is it acceptable to work on the top of a car without a car-top control station?

A  When the unit has been locked out and tagged

B  When two engineers are working on it

C  When there is no other way to work on it

D  When only one person is working on it

**21.39** Which statement is true when working on an energised car?

A  Always attach your lanyard to the car top while standing on the landing

B  Make sure you step onto the landing with your lanyard still attached to the car

C  Always check your lanyard is unclipped before getting off the car top

D  Ensure that your lanyard is clipped to a guide bracket or similar anchorage on the shaft

**21.40** What is the last thing you should do before getting off a car top through open landing doors when the car-top control is within 1 m of the landing threshold?

A  Set the car-top control to test

B  Ensure that the car-top stop button is set to stop and the car-top control remains set to test

C  Turn off the shaft lights and switch the car-top control to normal

D  Press the stop button and switch the car-top control to normal

**21.41** What precaution must you take if the landing doors are to be open while work goes on in the lift pit?

A  Erect a suitable barrier and secure it in front of the landing doors

B  Post a notice on the wall next to the doors

C  Do the job when there are not many people about

D  Ask someone to guard the open doors while you work

**21.42** What is the first thing to do after opening the landing doors and before accessing the car-top of an operating lift?

A  Chock the landing doors open

B  Press the car-top stop button

C  Make sure the lift has stopped

D  Put the car-top control in the test position

**21.43** When you gain access to a car top, how should you test that the car-top stop switch operates correctly?

A By trying to move the car in the up direction

B By trying to move the car in the down direction

C By measuring with a multimeter

D By flicking the switch on and off rapidly

**21.44** When working in the pit, when should the lift **not** be positioned towards the top of the shaft?

A When the hydraulic fluid level is low

B When the power supply is cut

C When you are testing the buffers

D When work needs to be done on the underside of the lift

**21.45** What is required on each landing of a new lift shaft before entrances and doors are fitted?

A A warning notice

B A substantial secure barrier to prevent falls

C Orange plastic netting across the opening

D Bright lighting

**21.46** When handling stainless steel car panels, which of the following items of personal protective equipment (PPE) should you wear in addition to safety footwear?

A Suitable safety gloves, such as rigger type gloves

B Hand barrier cream

C Latex gloves

D Hearing protection

**21.47** At what stage in the installation of a lift should guarding be fitted to the lift machine?

A At the end of the job

B During commissioning

C When handing over to the client

D Before the machine can be operated

**21.48** Which of these statements is **not** true?

A A stop switch must be within 1.5 m of the front of the car

B The car top should be clean and free from grease and oil spills

C You should secure your tools out of your standing area when working on top of the car

D Before trying to access the hoist way, you should decide whether the work will need the power supply to be live

**21.49** What is the most effective way of reducing the likelihood of being struck by falling objects?

A  Don't work below another person

B  Wear a safety helmet with a chin strap

C  Put warning signs up in the area

D  Install suitable debris netting

## 22 Tunnelling

**22.01** When working underground how quickly should you be able to get to your self-rescue set?

A It must be within one minute's walking distance

B It must be immediately available

C It must be within three minutes' walking distance

D It must be within two minutes' walking distance

**22.02** In the event of an emergency who will use the information displayed on a tally board?

A The Health and Safety Executive (HSE)

B The crane operator

C The rescue services

D Environmental Health

**22.03** If oxygen levels are dropping, at what point would the atmosphere be classed as oxygen deficient?

A 18%

B 19%

C 20%

D 21%

**22.04** Methane gas does not have an odour. In what two ways can it be dangerous?

A It causes skin irritation

B It can cause temporary blindness

C It is explosive

D It is toxic

E It reduces oxygen in the atmosphere

**22.05** Which of the following is a reliable way to detect carbon monoxide and methane gas?

A They both have a distinctive bad egg smell

B With a dosimeter

C With a calibrated gas detector

D With your eyes, as you can see the gases in the mist

**22.06** How could hydrogen sulphide affect those working in a tunnel?

A It can leave a yellow dust which can irritate the skin

B It can make it noisier and therefore harder for you to hear

C It can cause a mist making it difficult to see

D It can cause respiratory paralysis, stopping you from breathing

**22.07** How would you know if the ventilation system stops working?

A  You would be told at the daily start of shift briefing

B  You would be informed as part of your induction

C  An audible alarm would sound

D  Your supervisor would inform you

**22.08** Nitrogen oxide (NO$_x$) gas can be present in tunnels. Which of these plant items causes the most nitrogen oxide to be generated?

A  Tunnel-boring machines (TBMs)

B  Electro-hydraulic spray pumps

C  Rail-mounted plant

D  Diesel-powered equipment

**22.09** Exposure to nitrogen oxide (NO$_x$) gas can cause breathing problems. Which of the following should be the first control measure?

A  Provide air monitoring

B  Have a portable breathing set nearby

C  Avoid exposure

D  Provide ventilation systems

**22.10** How should communication equipment be powered?

A  Linked to the main tunnel power supply

B  Independent of the main tunnel power supply

C  Linked to the 33 kVA power supply

D  Battery-powered

**22.11** Which two methods are commonly used for communication between the tunnelling face and the surface?

A  Email

B  Two-way radio

C  Telephone

D  Text message

E  Tannoy system

**22.12** What is the maximum recommended distance between emergency lighting in a tunnel?

A  25 m

B  50 m

C  75 m

D  100 m

22

**22.13** What is the colour of a 400 volt plug?

A | Black

B | Blue

C | Yellow

D | Red

**22.14** When charging lead acid batteries they produce an explosive gas. Smoking and other naked flames are **not** permitted within what distance of a battery-charging area?

A | 5 m

B | 10 m

C | 15 m

D | 20 m

**22.15** Hot work activities are **not** permitted within what distance of a diesel-fuelling point?

A | 5 m

B | 10 m

C | 15 m

D | 20 m

**22.16** As a **minimum**, how long **must** a fire watch be maintained after hot works have been completed?

A | 15 minutes

B | 30 minutes

C | 45 minutes

D | 60 minutes

**22.17** What does the term **DCI** refer to in compressed air tunnelling?

A | Dizzy, confused, induced state

B | Decompression incident

C | Decompression, combustion incident

D | Decompression illness

**22.18** Which of the following characteristics is associated with hydrogen sulphide gas?

A | A smell of rotten eggs

B | It causes skin irritation

C | It is odourless

D | It causes a yellow haze

22

**22.19** Which of the following statements is correct regarding hand-arm vibration syndrome (HAVS)?

A  It can be prevented if gloves are worn

B  It can be partially cured with medication

C  It can be corrected by surgery

D  It causes irreversible damage

**22.20** What must be available at batching plants to deal with cement or concrete splashes?

A  A supply of running water

B  Sterile bandages

C  A first aider

D  An eyewash station

**22.21** Which one of the following is not a health risk associated with sprayed concrete linings?

A  Cement burns

B  Arc eye

C  Hand-arm vibration syndrome

D  Inhalation of dust

**22.22** Which of the following is not a hazard associated with hand mining?

A  Vibrating hand tools

B  Noise

C  Falling mined material

D  Sprayed concrete rebound

**22.23** When should a hop-up or refuge in the tunnel be used?

A  When vehicles are passing

B  For services such as cables

C  When installing ventilation cassettes

D  For storage of materials and equipment

**22.24** Which of the following must be fitted as a conveyor system safety device?

A  Emergency lighting

B  Emergency pull cord or stop button

C  Hazard lighting

D  Seat belt

22

**22.25** If a locomotive or vehicle is approaching you in the tunnel, when should you make your way to a hop-up or safe refuge?

- A | Immediately
- B | Only when you can see it
- C | When you have identified the direction of travel
- D | Only when everyone else starts to move

**22.26** A locomotive is entering the rear of the tunnel-boring machine. Which **two** electronic systems are recommended to assist in controlling its movements?

- A | Signal or traffic lights
- B | CCTV in the cab
- C | Siren
- D | Telephone
- E | Klaxon bell

**22.27** Inclined conveyors are fitted with anti-rollback devices to prevent the belt running backwards due to which **two** potential failures?

- A | Overloading
- B | Power loss
- C | Oil spillage
- D | Overheating
- E | Water leak

**22.28** Which of the following is a common traffic light system used underground to control plant movement?

- A | Red = stop, Amber = out bye, Green = in bye
- B | Red = stop, Amber = in bye, Green = out bye
- C | Red = in bye, Amber = stop, Green = out bye
- D | Red = out bye, Amber = in bye, Green = stop

**22.29** What is the likely hazard from moving plant or locomotives in a tunnel?

- A | Crush
- B | Crash
- C | Noise
- D | Asphyxiation

**22.30** Which one of the following is the **least** effective method of controlling locomotive movements in the pit bottom?

- A | Traffic lights
- B | Radio
- C | Shouting
- D | Hand signals

**22.31** How often should safe refuge (hop-ups) be located along a tunnel?

- A  50 m on straights; 25 m on curves
- B  60 m on straights; 30 m on curves
- C  70 m on straights; 25 m on curves
- D  80 m on straights; 20 m on curves

**22.32** The tunnel-boring machine operator has restricted vision during the building process. Which of the following is the most effective way of overcoming this?

- A  Alternative control point
- B  Mirrors at shoulder or crane level
- C  Use a signaller
- D  Get one of the gang to build

**22.33** What is the first action to be taken if there is a blockage in the grouting pipe?

- A  Locate the blockage
- B  Split the line
- C  Clean out section by section
- D  Release the pressure in the pipeline

**22.34** Why is it important to clean grouting pipelines after use?

- A  It helps prevent blockages, which could cause the hose to burst
- B  It helps prevent the pipes from becoming weakened
- C  It keeps the dust level below the point at which respiratory protective equipment (RPE) is needed
- D  It prevents the atmosphere from becoming explosive

**22.35** Crash cages and side bars or sliding doors should be fitted to all personnel carrying cars used in the tunnel. Select two reasons why they are needed.

- A  To prevent derailment of the personnel carrying car
- B  To minimise injury in the event of a derailment
- C  To allow personnel to talk while being transported
- D  To stop personnel sitting close together
- E  To prevent personnel leaning out or falling out

**22.36** For tunnelling operations what is the minimum number of escape routes or methods that must be maintained from a working shaft?

- A  One
- B  Two
- C  Three
- D  Four

22

**22.37** What term is used for the access and egress control system to tunnels?

A Visitor book

B Tally system

C Signing-in book

D Clocking-on machine

**22.38** If a personnel carrying cage is used to transport workers into and out of a shaft, how many people can be transported at any one time?

A As defined by the Health and Safety Executive (HSE)

B As many as can fit into it

C As many as is stated on the personnel carrier

D As many as is stated by the supervisor

**22.39** Which of the following situations would require using a safety harness?

A Working as the signaller at the pit top

B Working as the belt-person on a TBM

C Landing concrete jacking-pipes in the pit bottom

D Building rings from a platform within a shaft

**22.40** What is the minimum recommended height of a secure barrier used to prevent falls around an open shaft?

0.9 m      1.2 m      1.5 m      1.8 m

**22.41** Oxygen cylinders should not be allowed to come into contact with which of the following substances?

- A   Mud
- B   Grease
- C   Paint
- D   Air

**22.42** While walking through the tunnel you see a tear in the ventilation ducting. What should you do?

- A   Report it to your supervisor
- B   Try to repair it
- C   Check if it has got any bigger at the end of the shift
- D   Evacuate the tunnel

**22.43** A grout gun inserted into a segment grout hole may present which of the following hazards?

- A   Shrinking grout hose
- B   Falls from height
- C   Blowout at injection point
- D   Hand-arm vibration

**22.44** Which of the following personal protective equipment (PPE) is not normally required for robotic-sprayed concrete lining operations in tunnelling?

- A   Eye protection
- B   Respiratory protective equipment (RPE)
- C   Disposable overalls
- D   Safety harness

22

**23.01** When a new piece of plant has been installed but has **not** been commissioned, how should it be left?

A   With all valves and switches turned off

B   With all valves and switches clearly labelled

C   With all valves and switches locked off

D   With all valves and switches turned on and ready to use

**23.02** Who can solder a fitting on an isolated copper gas pipe?

A   A plumber

B   A pipefitter

C   A skilled welder

D   A Gas Safe registered engineer

**23.04** When working in a riser, how should access be controlled?

A   By a site security operative

B   By those who are working in it

C   By the main contractor

D   By a permit to work system

**23.04** If you find a coloured wire sticking out of an electrical plug what is the correct action to take?

A   Push it back into the plug and carry on working

B   Pull the wire clear of the plug and report it to your supervisor

C   Mark the item as defective and follow your company procedure for defective items

D   Take the plug apart and carry out a repair

**23.05** How should extension leads in use on site be positioned?

A   They should be located so as to prevent a tripping hazard

B   They should be laid out in the shortest, most convenient route

C   They should be coiled on a drum or cable tidy

D   They should be raised on bricks

**23.06** What should you do if you need additional temporary wiring for your power tools whilst working on site?

A   Find some cable and extend the wiring yourself

B   Stop work until an authorised supply has been installed

C   Speak to an electrician and ask them to do the temporary wiring

D   Disconnect a longer cable serving somewhere else and reconnect it to where you need it

**23.07** Which item of electrical equipment does not require portable appliance testing?

A  Battery-powered rechargeable drill

B  110 volt electrical drill

C  110 volt portable halogen light

D  Electric kettle

**23.08** Why is temporary continuity bonding carried out before removing and replacing sections of metallic pipework?

A  To provide a continuous earth for the pipework installation

B  To prevent any chance of blowing a fuse

C  To maintain the live supply to the electrical circuit

D  To prevent any chance of corrosion to the pipework

**23.09** Which type of power drill is most suitable for fixing a run of pipework outside in wet weather?

A  Battery-powered drill

B  Drill with 110 volt power supply

C  Drill with 240 volt power supply

D  Any mains voltage drill with a power breaker

**23.10** What would you use to find out whether a wall into which you are about to drill contains an electric supply?

A  A neon screwdriver

B  A cable tracer

C  A multimeter

D  A hammer and chisel

**23.11** Where should liquefied petroleum gas (LPG) cylinders be positioned when supplying an appliance in a site cabin?

A  Inside the cabin in a locked cupboard

B  Under the cabin

C  Inside the cabin next to the appliance

D  Outside the cabin

**23.12** How should you position the exhaust of an engine-driven generator that has to be run inside a building?

A  Outside the building

B  In a stairwell

C  In another room

D  In a riser

23

**23.13** How should cylinders containing liquefied petroleum gas (LPG) be stored on site?

A In a locked cellar with clear warning signs

B In a locked cage at least 3 m from any oxygen cylinders

C Within a secure storage container at the back of the site

D Covered by a tarpaulin to shield the compressed cylinder from sunlight

**23.14** If you spill some oil on the floor and you do **not** have any absorbent material to clean the area, what should you do?

A Spread it about to lessen the depth

B Keep people out of the area and inform your supervisor

C Do nothing, as it will eventually soak into the floor

D Warn other people as they tread through it

**23.15** When **must** you wear eye protection while drilling through a wall?

A Only when drilling overhead

B Only when the drill bit exceeds 20 mm

C Always, whatever the circumstances

D Only when drilling through concrete

**23.16** What personal protective equipment (PPE) should you wear when using a hammer drill to drill a 100 mm diameter hole through a brick wall?

A Gloves, breathing apparatus and boots

B Ear defenders, respiratory protective equipment, boots and eye protection

C Ear defenders, breathing apparatus and barrier cream

D Barrier cream, boots and respiratory protective equipment

**23.17** What should you do when using pipe-freezing equipment to isolate a damaged section of pipe?

A Always work in pairs when using pipe-freezing equipment

B Never allow the freezing gas to come into direct contact with surface water

C Never use pipe-freezing equipment on plastic pipe

D Wear gloves to avoid direct contact with the skin and read the COSHH assessment

**23.18** If you find a dangerous gas fitting that is likely to cause a death or specified injury, who **must** be sent a formal report?

A The client

B The gas board

C The health and safety manager

D The Health and Safety Executive (HSE)

**23.19** If you are working where welding is being carried out, what should be provided to protect you from welding flash?

- A  A fire extinguisher
- B  Warning notices
- C  Screens
- D  A hi-vis vest

**23.20** When using a blowtorch to joint copper tube and fittings in a domestic property, how should a fire extinguisher be made available?

- A  It should be available in the immediate work area
- B  It should be held over the joint while you are using the blowtorch
- C  It should be kept away from the work area in case a spark causes it to explode
- D  It should be available only if a property is occupied

**23.21** If you are carrying out hot works with a blowtorch, when should you stop using it?

- A  Just before you leave the site
- B  At least one hour before you leave the site
- C  At least two hours before you leave the site
- D  At least four hours before you leave the site

**23.22** What should you do when using a blowtorch near to flexible pipe lagging?

- A  Remove the lagging at least 1 m either side of the work
- B  Remove just enough lagging to carry out the work
- C  Remove the lagging at least 3 m either side of the work
- D  Wet the lagging but leave it in place

**23.23** What should you do when using a blowtorch near to timber?

- A  Carry out the work, taking care not to set fire to the timber
- B  Wet the timber first and have a bucket of water handy
- C  Use a non-combustible mat and have a fire extinguisher in the immediate work area
- D  Point the flame away from the timber and have a bucket of sand ready to put out the fire

**23.24** The legionella bacteria that cause legionnaires' disease are most likely to be found in which of the following?

- A  A boiler operating at a temperature of 80°C
- B  An infrequently used shower hose outlet
- C  A cold water storage cistern containing water at 10°C
- D  A toilet pan

23

# SPECIALIST

**23.25** How are legionella bacteria passed on to humans?

A. Through fine water droplets, such as sprays or mists

B. By drinking dirty water

C. Through contact with the skin

D. From other people when they sneeze

**23.26** When planning a lifting operation, how should the sequence of operations to enable a safe lift be confirmed?

A. Using verbal instruction

B. In a method statement

C. In a radio telephone message

D. On a notice in the canteen

**23.27** Which of these is true in relation to the safe working load (SWL) of lifting equipment?

A. It is never marked on the equipment but kept with the test certificates

B. It is provided for guidance only

C. It may be exceeded by no more than 25%

D. It is the maximum safe working load

**23.28** What must be clearly marked on all lifting equipment?

A. The name of the manufacturer

B. The safe working load

C. The next test date

D. The specification of material from which it is made

**23.29** When carrying a ladder on a vehicle, what is the correct way of securing the ladder to the roof rack?

A. Rope

B. Bungee elastics

C. Ladder clamps

D. Copper wire

**23.30** What is the safest method of transporting long lengths of copper pipe by van?

A. Tying the pipes to the roof with copper wire

B. Someone holding the pipes on the roof rack as you drive along

C. Putting the pipes inside the van with the ends out of the passenger window

D. Using a pipe rack fixed to the roof of the van

**23.31** What is the safest way to move a cast iron boiler some distance?

A · Get a workmate to carry it with you

B · Drag it

C · Roll it end-over-end

D · Use a trolley or other manual handling aid

**23.32** During a job you may need to work below a ground-level suspended timber floor. What is the most important question you should ask?

A · Can the work be performed from outside?

B · Will temporary lighting be used?

C · How many days will the work take to complete?

D · Could Weil's disease (leptospirosis) be a problem?

**23.33** What must you ensure before using a ladder?

A · That it is secured to prevent it from moving sideways or sliding outwards

B · That no-one else has booked the ladder for their work

C · That an apprentice or workmate is standing by in case you slip and fall

D · That the weather forecast is for a bright, clear day

**23.34** When positioning and erecting a stepladder, which of the following is essential for its safe use?

A · It has a tool tray towards the top of the steps

B · The restraint mechanism is spread to its full extent

C · You will be able to reach the job by standing on the top step

D · Your supervisor has positioned and erected the steps

**23.35** What is the recommended maximum height for a free-standing mobile tower?

A · There is no recommended height

B · 2 m

C · The height recommended by the manufacturer

D · 12 m

**23.36** What is the first thing you should do after getting on to the platform of a correctly erected mobile tower?

A · Check that the brakes are locked on

B · Check for overhead power lines

C · Check that the access hatch has been closed to prevent falls of personnel, tools or equipment

D · Check that the tower does not rock or wobble

23

**23.37** What should be done before a mobile tower is moved?

A All people and equipment must be removed from the platform

B A permit to work must be issued

C The principal contractor must give their approval

D Arrangements must be made with the forklift truck driver

**23.38** What must be done **first** before any roof work is carried out?

A A risk assessment must be carried out

B The operatives working on the roof must be trained in the use of safety harnesses

C Permits to work must be issued only to those allowed to work on the roof

D A weather forecast must be obtained

**23.39** What is edge protection designed to do?

A Make access to the roof easier

B Secure tools and materials close to the edge

C Prevent rainwater running off the roof onto workers below

D Prevent the fall of people and materials

**23.40** When assembling a mobile tower what major hazard **must** you be aware of?

A Water pipes

B Cable trays

C False ceilings

D Overhead service cables

**23.41** What should folding stepladders be used for?

A General access on site

B Short-term work

C All site activities where a straight ladder cannot be used

D Getting on and off mobile towers

**23.42** How should you access a roof to install a flexible flue liner into an existing chimney?

A Work from a roof ladder securely hooked over the ridge

B Use an access scaffold designed for chimney works

C Scramble up the roof tiles to get to the chimney

D Get your mate to do the job while you hold a rope tied to them

23

**23.43** When drilling a hole for a boiler flue outside, which type of working platform should you use?

A A long ladder

B Borrowed scaffolding that you have erected

C A mobile tower or fixed scaffold

D Packing cases to stand on

**23.44** What is the only circumstance where stepladders should be used?

A Inside buildings

B If no other suitable equipment is available after works have been risk assessed

C If they are made of aluminium

D If they are less than 1.75 m high

## 24    HVACR – Pipefitting and welding

**24.01** When a new piece of plant has been installed but has **not** been commissioned, how should it be left?

- A  With all valves and switches turned off
- B  With all valves and switches clearly labelled
- C  With all valves and switches locked off
- D  With all valves and switches turned on and ready to use

**24.02** Who is allowed to install natural gas pipework?

- A  A skilled engineer
- B  A pipefitter
- C  A Gas Safe registered engineer
- D  Anybody

**24.03** Who should carry out pressure testing on pipework or vessels?

- A  Anyone who is available
- B  A competent person
- C  A Health and Safety Executive (HSE) inspector
- D  A building control officer

**24.04** When working in a riser, how should access be controlled?

- A  By a site security operative
- B  By those who are working in it
- C  By the main contractor
- D  By a permit to work system

**24.05** While working on your own and tracing pipework in a building, you notice that the pipes enter a service duct. What should you do?

- A  Go into the service duct and continue to trace the pipework
- B  Ask someone in the building to act as your second person
- C  Put on your personal protective equipment (PPE) and carry on with the job
- D  Stop work until a risk assessment has been carried out

**24.06** If you find a coloured wire sticking out of an electrical plug what is the correct action to take?

- A  Push it back into the plug and carry on working
- B  Pull the wire clear of the plug and report it to your supervisor
- C  Mark the item as defective and follow your company procedure for defective items
- D  Take the plug apart and carry out a repair

**24.07** What should you do if you need additional temporary wiring for your power tools whilst working on site?

A — Find some cable and extend the wiring yourself

B — Stop work until an authorised supply has been installed

C — Speak to an electrician and ask them to do the temporary wiring

D — Disconnect a longer cable serving somewhere else and reconnect it to where you need it

**24.08** Which item of electrical equipment does not require portable appliance testing?

A — Battery-powered rechargeable drill

B — 110 volt electrical drill

C — 110 volt portable halogen light

D — Electric kettle

**24.09** How should extension leads in use on site be positioned?

A — They should be located so as to prevent a tripping hazard

B — They should be laid out in the shortest, most convenient route

C — They should be coiled on a drum or cable tidy

D — They should be raised on bricks

**24.10** Where should liquefied petroleum gas (LPG) cylinders be positioned when supplying an appliance in a site cabin?

A — Inside the cabin in a locked cupboard

B — Under the cabin

C — Inside the cabin next to the appliance

D — Outside the cabin

**24.11** How should you position the exhaust of an engine-driven generator that has to be run inside a building?

A — Outside the building

B — In a stairwell

C — In another room

D — In a riser

**24.12** How should cylinders containing liquefied petroleum gas (LPG) be stored on site?

A — In a locked cellar with clear warning signs

B — In a locked cage at least 3 m from any oxygen cylinders

C — As close to the point of use as possible

D — Covered by a tarpaulin to shield the compressed cylinder from sunlight

24

**24.13** If you spill some oil on the floor and you do **not** have any absorbent material to clean the area, what should you do?

A Spread it about to lessen the depth

B Keep people out of the area and inform your supervisor

C Do nothing, as it will eventually soak into the floor

D Warn other people as they tread through it

**24.14** What should you do when using pipe-freezing equipment to isolate a damaged section of pipe?

A Always work in pairs

B Never allow the freezing gas to come into direct contact with surface water

C Never use pipe-freezing equipment on plastic pipe

D Wear gloves to avoid direct contact with your skin and read the COSHH assessment

**24.15** Why is it important to know the difference between propane and butane equipment?

A Propane equipment operates at higher pressure

B Propane equipment operates at lower pressure

C Propane equipment is cheaper

D Propane equipment can be used with smaller, easy-to-handle cylinders

**24.17** Which of the following statements is **true**?

A Both propane and butane are heavier than air

B Butane is heavier than air while propane is lighter than air

C Propane is heavier than air while butane is lighter than air

D Both propane and butane are lighter than air

**24.17** Apart from the cylinders used in gas-powered forklift trucks, why should liquefied petroleum gas (LPG) cylinders **never** be placed on their side during use?

A It would give a faulty reading on the contents gauge, resulting in flashback

B Air could be drawn into the cylinder, creating a dangerous mixture of gases

C The liquid gas would be at too low a level to allow the torch to burn correctly

D The liquid gas could be drawn from the cylinder, creating a safety hazard

**24.18** What is the method of checking for leaks after connecting a liquefied petroleum gas (LPG) regulator to the bottle?

A Test with a lighted match

B Sniff the connections to detect the smell of gas

C Listen to hear for escaping gas

D Apply leak detection fluid to the connections

24

**24.19** What is the **most** likely risk of injury when cutting a pipe with hand-operated pipe cutters?

A Your fingers may become trapped between the cutting wheel and the pipe

B You may cut yourself on the inside edge of the cut pipe

C You may damage your muscles due to continued use

D A piece of sharp metal could fly off and hit you

**24.20** Why is it essential to take great care when handling oxygen cylinders?

A They contain highly flammable compressed gas

B They contain highly flammable liquid gas

C They are filled to extremely high pressures

D They contain poisonous gas

**24.21** When do you need to wear eye protection while drilling through a wall?

A Only when drilling overhead

B Only when the drill bit exceeds 20 mm

C Always, whatever the circumstances

D Only when drilling through concrete

**24.22** What guarding is required when a pipe threading machine is in use?

A A length of red material hung from the exposed end of the pipe

B A barrier at the exposed end of the pipe only

C A barrier around the whole of the pipe length and machine

D Warning notices in the work area

**24.23** If you are working where welding is being carried out, what should be provided to protect you from welding flash?

A A fire extinguisher

B Warning notices

C Screens

D A hi-vis vest

**24.24** What is the main hazard associated with flame-cutting and welding?

A Gas poisoning

B Fire

C Dropping a gas cylinder

D Not having a hot-work permit

24

**24.25** When should you stop carrying out hot works?

A Just before you leave the site

B At least one hour before you leave the site

C At least two hours before you leave the site

D At least four hours before you leave the site

**24.26** What should you do when using a blowtorch near to flexible pipe lagging?

A Remove the lagging at least 1 m either side of the work

B Remove just enough lagging to carry out the work

C Remove the lagging at least 3 m either side of the work

D Wet the lagging but leave it in place

**24.27** When using a blowtorch near to timber, what should you do?

A Carry out the work taking care not to catch the timber

B Point the flame away from the timber and have a bucket of sand ready to put out the fire

C Wet the timber first and keep a bucket of water handy

D Use a non-combustible mat and have a fire extinguisher ready

**24.28** What is the colour of an acetylene cylinder?

A Orange

B Black

C Green

D Maroon

**24.29** Which item of personal protective equipment (PPE) is designed to protect against infrared radiation damage to the eyes during flame cutting or welding?

A Respiratory protective equipment (RPE)

B Clear goggles

C Eye protection with a tinted filter lens

D Dust mask

**24.30** When using oxyacetylene brazing equipment, how should the bottles be positioned?

A Laid on their side and secured

B Stood upright and secured

C Stood upside down

D Angled at 45°

**24.31** The use of oxyacetylene equipment is **not** recommended for which jointing method?

A   Jointing copper pipe using hard soldering

B   Jointing copper tube using capillary soldered fittings

C   Jointing mild steel tube

D   Jointing sheet lead

**24.32** Where should acetylene gas-welding bottles be stored when they are **not** in use?

A   Outside in a special storage compound

B   In a special rack in a company van

C   Inside a building in a locked cupboard

D   With oxygen bottles

**24.34** When planning a lifting operation, how should the sequence of operations to enable a safe lift be confirmed?

A   Using verbal instruction

B   In a method statement

C   In a radio telephone message

D   On a notice in the canteen

**24.35** Which of these statements is **true** in relation to the safe working load (SWL) of lifting equipment?

A   It is never marked on the equipment but kept with the test certificates

B   It is provided for guidance only

C   It may be exceeded by no more than 25%

D   It is the absolute maximum safe working load

**24.33** Which **two** of these activities are likely to need a hot-work permit?

You will be asked to 'drag and drop' your answers

A   Refueling a diesel dump truck

B   Using the heaters in the drying room

C   Cutting steel with an angle grinder

D   Soldering pipework in a central heating system

E   Replacing an empty liquefied petroleum gas (LPG) cylinder with a full one

24

**24.36** What **must** be clearly marked on all lifting equipment?

A. The name of the manufacturer

B. The safe working load

C. The next test date

D. The specification of material from which it is made

**24.37** Which **two** of the following are essential safety checks to be carried out before using oxyacetylene equipment?

A. That the cylinders are full

B. That the cylinders, hoses and flashback arresters are in good condition

C. That the trolley wheels are the right size

D. That the area is well ventilated and clear of any obstructions

E. That the cylinders are the right weight

**24.38** Who should be present during the pressure testing of pipework or vessels?

A. The architect

B. The site foreman

C. Only those involved in carrying out the test

D. Anybody

**24.39** What should you ensure when using an electrically powered threading machine?

A. That the power supply is 24 volts

B. That the power supply is 400 volts and the machine is fitted with a guard

C. That your clothing cannot get caught on rotating parts of the machine

D. That the machine is only used in your compound

**24.40** What **must** you ensure before using a ladder?

A. That it is secured to prevent it from moving sideways or sliding outwards

B. That no-one else has booked the ladder for their work

C. That an apprentice or workmate is standing by in case you slip and fall

D. That the weather forecast is for a bright, clear day

**24.41** When positioning and erecting a stepladder, which of the following is essential for its safe use?

A. It has a tool tray towards the top of the steps

B. The restraint mechanism is spread to its full extent

C. You will be able to reach the job by standing on the top step

D. Your supervisor has positioned and erected the steps

24

**24.42** What is the recommended **maximum** height for a free-standing mobile tower?

A There is no restriction

B 2 m

C The height recommended by the manufacturer

D 12 m

**24.43** What should be done before a mobile tower is moved?

A All people and equipment must be removed from the platform

B A permit to work must be issued

C The principal contractor must give their approval

D Arrangements must be made with the forklift truck driver

**24.44** What must be done **first** before any roof work is carried out?

A A risk assessment must be carried out

B The operatives working on the roof must be trained in the use of safety harnesses

C Permits to work must be issued only to those allowed to work on the roof

D A weather forecast must be obtained

**24.45** What is edge protection designed to do?

A Make access to the roof easier

B Secure tools and materials close to the edge

C Prevent rainwater running off the roof onto workers below

D Prevent the fall of people and materials

**24.46** What is the **first** thing you should do after getting on to the platform of a correctly erected mobile tower?

A Check that the brakes are locked on

B Check for overhead power lines

C Close the access hatch to prevent falls of personnel, tools or equipment

D Make sure that the tower does not rock or wobble

**24.47** When assembling a mobile tower what major hazard **must** you be aware of?

A Water pipes

B Cable trays

C False ceilings

D Overhead service cables

24

**24.48** 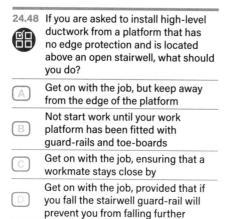 If you are asked to install high-level ductwork from a platform that has no edge protection and is located above an open stairwell, what should you do?

| A | Get on with the job, but keep away from the edge of the platform |

| B | Not start work until your work platform has been fitted with guard-rails and toe-boards |

| C | Get on with the job, ensuring that a workmate stays close by |

| D | Get on with the job, provided that if you fall the stairwell guard-rail will prevent you from falling further |

24

## 25    HVACR – Ductwork

**25.01** When a new piece of plant has been installed but has **not** been commissioned, how should it be left?

- A  With all valves and switches turned off
- B  With all valves and switches clearly labelled
- C  With all valves and switches locked off
- D  With all valves and switches turned on and ready to use

**25.02** Who should carry out leakage testing of a newly installed ductwork system?

- A  The installation contractor
- B  The property owner
- C  The designer
- D  A trained and competent person

**25.03** When working in a riser, how should access be controlled?

- A  By a site security operative
- B  By those who are working in it
- C  By the main contractor
- D  By a permit to work system

**25.04** If you find a coloured wire sticking out of an electrical plug what is the correct action to take?

- A  Push it back into the plug and carry on working
- B  Pull the wire clear of the plug and report it to your supervisor
- C  Mark the item as defective and follow your company procedure for defective items
- D  Take the plug apart and carry out a repair

**25.05** How should extension leads in use on site be positioned?

- A  They should be located so as to prevent a tripping hazard
- B  They should be laid out in the shortest, most convenient route
- C  They should be coiled on a drum or cable tidy
- D  They should be raised on bricks

**25.06** What should you do if you need additional temporary wiring for your power tools whilst working on site?

- A  Find some cable and extend the wiring yourself
- B  Stop work until an authorised supply has been installed
- C  Speak to an electrician and ask them to do the temporary wiring
- D  Disconnect a longer cable serving somewhere else and reconnect it to where you need it

25

**25.07** Which item of electrical equipment does **not** require portable appliance testing?

A Battery-powered rechargeable drill

B 110 volt electrical drill

C 110 volt portable halogen light

D Electric kettle

**25.08** Where should liquefied petroleum gas (LPG) cylinders be positioned when supplying an appliance in a site cabin?

A Inside the cabin in a locked cupboard

B Under the cabin

C Inside the cabin next to the appliance

D Outside the cabin

**25.09** How should you position the exhaust of an engine-driven generator that has to be run inside a building?

A Outside the building

B In a stairwell

C In another room

D In a riser

**25.10** If you spill some oil on the floor and you do **not** have any absorbent material to clean the area, what should you do?

A Spread it about to lessen the depth

B Keep people out of the area and inform your supervisor

C Do nothing, as it will eventually soak into the floor

D Warn other people as they tread through it

**25.11** A person who has been using a solvent-based ductwork sealant is complaining of headaches and feeling sick. What is the **first** thing you should do?

A Let them carry on working but try to keep a close watch on them

B Get them a drink of water and a headache tablet

C Get them out to fresh air and make them rest

D Make an entry in the accident book

**25.12** What additional control measure **must** be put in place when welding in-situ galvanised ductwork?

A Screens

B Fume extraction

C Warning signs

D Hearing protection

25

**25.13** When jointing plastic-coated metal ductwork, which of the following methods of jointing presents the **most** serious risk to health?

- A Welding
- B Taping
- C Riveting
- D Fixing nuts and bolts

**25.14** If you are removing a run of ductwork in an unoccupied building and notice a hypodermic syringe behind it, what should you do?

- A Ensure the syringe is empty, remove it and place it with the rubbish
- B Wear gloves, break the syringe into small pieces and flush it down the drain
- C Notify the supervisor, cordon off the area and call the emergency services
- D Wear gloves, use grips to remove the syringe to a safe place and report your find

**25.15** Which two of the following should you use when cutting aluminium or tin ductwork that has been pre-insulated with fibreglass?

- A A hacksaw
- B Respiratory protective equipment
- C A chisel
- D A set of tin snips
- E A blowtorch

**25.16** If you are working where welding is being carried out, what should be provided to protect you from welding flash?

- A A fire extinguisher
- B Warning notices
- C Screens
- D A hi-vis vest

**25.17** When planning a lifting operation, how should the sequence of operations to enable a safe lift be confirmed?

- A By verbal instructions
- B In a method statement
- C In a radio telephone message
- D Via a notice in the canteen

**25.18** Which of these statements is true in relation to the safe working load (SWL) of lifting equipment?

- A It is never marked on the equipment but kept with the test certificates
- B It is provided for guidance only
- C It may be exceeded by no more than 25%
- D It is the absolute maximum safe working load

25

**25.19** What **must** be clearly marked on all lifting equipment?

A. The name of the manufacturer

B. The safe working load

C. The next test date

D. The specification of material from which it is made

**25.20** When using a material hoist you notice that the lifting cable is frayed. What should you do?

A. Get the job done as quickly as possible

B. Straighten out the cable using mole grips

C. Do not use the hoist, and report the problem

D. Be very careful when using the hoist

**25.21** When do you need to wear eye protection while drilling through a wall?

A. Only when drilling overhead

B. Only when the drill bit exceeds 20 mm

C. Always, whatever the circumstances

D. Only when drilling through concrete

**25.22** In addition to a safety helmet and protective footwear, what personal protective equipment (PPE) should you wear when using a hammer drill?

A. Gloves and breathing apparatus

B. Hearing protection, respiratory protective equipment and eye protection

C. Hearing protection, breathing apparatus and barrier cream

D. Barrier cream and respiratory protective equipment

**25.23** How should you leave the area around ductwork after using a solvent-based sealant?

A. Seal up all the open ends to ensure that dirt cannot get into the system

B. Ensure that the lids are left off tins of solvent

C. Remove any safety signs or notices

D. Leave inspection covers off and erect no smoking signs

**25.24** Before taking down a run of ductwork, what is the **first** thing you should do?

A. Assess the volume of waste and get an appropriate sized skip

B. Cut through the support rods

C. Clean the ductwork to remove all dust

D. Assess the task to be undertaken and check its support system

**25.25** What is the safest way to move a fan-coil unit some distance?

A. Get a workmate to carry it with you

B. Drag it

C. Roll it end-over-end

D. Use a trolley or other manual handling aid

**25.26** While fitting a fire damper into a ductwork system you notice that, due to a manufacturing fault, it may not operate properly. What should you do?

A. Install it anyway, as it is

B. Fix it so that it stays open, and then install it

C. Not fit the damper and report the fault

D. Leave it out of the ductwork system altogether

**25.27** If you are using a genie hoist and notice that part of the hoist is buckling slightly, what should you do?

A. Lower the load immediately

B. Carry on with the job, while keeping an eye on the buckling metal

C. Straighten out the buckled metal and then get on with the lifting operation

D. Get the job finished quickly

**25.28** When carrying out solvent welding on plastic ductwork, what particular safety measure must be applied?

A. The area must be well ventilated

B. The supervisor must be present

C. A hard hat must be worn

D. It must be done in daylight

**25.29** Which of the following do you not need to do before using a cleaning agent or biocide in a ductwork system?

A. Ask for advice from the cleaning agent or biocide manufacturer

B. Read the COSHH assessment for the material, carry out a risk assessment and produce a method statement for the work

C. Consult the building occupier

D. Check what the ductwork will carry in the future

**25.30** Which of the following do you not need to do before cleaning a system in industrial, laboratory or other premises where you might encounter harmful particulates?

A. Examine the system

B. Collect a sample from the ductwork

C. Run the system under overload conditions

D. Prepare a job-specific risk assessment and method statement

25

**25.31** Where it is necessary to enter ductwork, which are the **two** main factors that need to be considered?

- A   Working in a confined space
- B   What the ductwork will carry in the future
- C   The cleanliness of the ductwork
- D   Whether you need to wear kneepads
- E   The strength of the ductwork and its supports

**25.32** What should you do before painting the external surface of ductwork?

- A   Clean the paintbrushes
- B   Read the COSHH assessment
- C   Switch off the system
- D   Put on eye protection

**25.33** If you have to dismantle some waste-extract ductwork, what is the **first** thing you should do?

- A   Arrange for a skip to put it in
- B   Ensure there is a certificate of cleanliness in place before commencing work
- C   Check that the duct supports are strong enough to cope with the dismantling
- D   Make sure there are enough disc cutters to do the job

**25.34** What **must** you ensure before using a ladder?

- A   That it is secured to prevent it from moving sideways or sliding outwards
- B   That no-one else has booked the ladder for their work
- C   That an apprentice or workmate is standing by in case you slip and fall
- D   That the weather forecast is for a bright, clear day

**25.35** When positioning and erecting a stepladder, which of the following is essential for its safe use?

- A   It has a tool tray towards the top of the steps
- B   The restraint mechanism is spread to its full extent
- C   You will be able to reach the job by standing on the top step
- D   Your supervisor has positioned and erected the steps

**25.36** What is the recommended **maximum** height for a free-standing mobile tower?

- A   There is no restriction
- B   2 m
- C   The height recommended by the manufacturer
- D   12 m

**25.37** What is the **first** thing you should do after getting on to the platform of a correctly erected mobile tower?

A  Check that the brakes are locked on

B  Check for overhead power lines

C  Close the access hatch to prevent falls of personnel, tools or equipment

D  Make sure that the tower does not rock or wobble

**25.38** What should be done before a mobile tower is moved?

A  All people and equipment must be removed from the platform

B  A permit to work must be issued

C  The principal contractor must give their approval

D  Arrangements must be made with the forklift truck driver

**25.39** What must be done **first** before any roof work is carried out?

A  A risk assessment must be carried out

B  The operatives working on the roof must be trained in the use of safety harnesses

C  Permits to work must be issued only to those allowed to work on the roof

D  A weather forecast must be obtained

**25.40** What is edge protection designed to do?

A  Make access to the roof easier

B  Secure tools and materials close to the edge

C  Prevent rainwater running off the roof onto workers below

D  Prevent the fall of people and materials

**25.41** What should folding stepladders be used for?

A  General access on site

B  Short-term works

C  All site activities where a straight ladder cannot be used

D  Getting on and off mobile towers

**25.42** If you are asked to install high-level ductwork from a platform that has no edge protection and is located above an open stairwell, what should you do?

A  Get on with the job, but keep away from the edge of the platform

B  Not start work until your work platform has been fitted with guard-rails and toe-boards

C  Get on with the job, ensuring that a workmate stays close by

D  Get on with the job, provided that if you fall the stairwell guard-rail will prevent you from falling further

25

**25.43** You have to carry out a job over a few days on the flat roof of a two-storey building, about 1 m from the edge of the roof, which has a low parapet. How should you reduce your risk of falling?

A  Carry on with the job, provided that you don't get dizzy with heights

B  Use a full body harness, lanyard and anchor while doing the job

C  Ask for double guard-rails and a toe-board to be installed

D  Get your mate to do the work, while you hold on to them

**25.44** What is the best form of access to use, when installing a run of ceiling-mounted ductwork across a large open space?

A  Stepladder

B  Metal trestles

C  Wooden trestles

D  Mobile tower

**25.45** When assembling a mobile tower what major hazard must you be aware of?

A  Water pipes

B  Cable trays

C  False ceilings

D  Overhead service cables

# 26 HVACR – Refrigeration and air conditioning

**26.01** When a new piece of plant has been installed but has **not** been commissioned, how should it be left?

A  With all valves and switches turned off

B  With all valves and switches clearly labelled

C  With all valves and switches locked off

D  With all valves and switches turned on and ready to use

**26.02** When working on refrigeration systems containing hydrocarbon (HC) gases, what particular danger needs to be considered?

A  There should be no sources of ignition

B  Special personal protective equipment (PPE) should be worn to prevent injuries caused by the cold

C  Extra lighting is needed to prevent trips

D  The work cannot be carried out when the weather is hot

**26.03** What should you do when it is necessary to cut into an existing refrigerant pipe?

A  Vent the gas in the pipework to atmosphere

B  Recover the refrigerant gas and make a record of it, then do the work

C  Work on the pipework with the refrigerant gas still in it

D  Not carry out the work at all, because of the risks

**26.04** What is the **first** thing that should be done when a new refrigeration system has been installed?

A  It should be pressure and leak tested

B  It should be filled with refrigerant

C  It should be left open to the air

D  It should be turned off at the electrical switch

**26.05** Which of these statements is **true** of the water in water-cooled systems?

A  It should be replaced annually

B  It should be chemically treated

C  It should be properly filtered

D  It should be drinking water

**26.06** Who is permitted to install, service or maintain systems that contain or are designed to contain refrigerant gases?

A  A Gas Safe registered engineer

B  The person whose plant contains the gas

C  A competent, trained person who works for an F-Gas registered company

D  A fully qualified electrician

26

**26.07** When working in a riser, how should access be controlled?

A By a site security operative

B By those who are working in it

C By the main contractor

D By a permit to work system

**26.08** If you find a coloured wire sticking out of an electrical plug what is the correct action to take?

A Push it back into the plug and carry on working

B Pull the wire clear of the plug and report it to your supervisor

C Mark the item as defective and follow your company procedure for defective items

D Take the plug apart and carry out a repair

**26.09** How should extension leads in use on site be positioned?

A They should be located so as to prevent a tripping hazard

B They should be laid out in the shortest, most convenient route

C They should be coiled on a drum or cable tidy

D They should be raised on bricks

**26.10** What should you do if you need additional temporary wiring for your power tools whilst working on site?

A Find some cable and extend the wiring yourself

B Stop work until an authorised supply has been installed

C Speak to an electrician and ask them to do the temporary wiring

D Disconnect a longer cable serving somewhere else and reconnect it to where you need it

**26.11** Which item of electrical equipment does **not** require portable appliance testing?

A 110 volt electrical power tool

B Battery-powered rechargeable power tool

C 240 volt electrical power tool

D 240 volt charger for battery-powered tools

**26.12** When repairing an electrically driven compressor, what is the **minimum** safe method of isolation?

A Pressing the stop button

B Pressing the emergency stop button

C Turning off the local isolator

D Locking off and tagging out the local isolator

26

**26.13** When a refrigerant leak is reported in a closed area, what should you do **first** before entering the area?

- A Ventilate the area
- B Establish that it is safe to enter
- C Get a torch
- D Wear safety footwear

**26.14** Where should liquefied petroleum gas (LPG) cylinders be positioned when supplying an appliance in a site cabin?

- A Inside the cabin in a locked cupboard
- B Under the cabin
- C Inside the cabin next to the appliance
- D Outside the cabin

**26.15** How should you position the exhaust of an engine-driven generator that has to be run inside a building?

- A Outside the building
- B In a stairwell
- C In another room
- D In a riser

**26.16** How should cylinders (full or empty) that contain liquefied petroleum gas (LPG) or acetylene be stored on site?

- A In a locked cellar with clear warning signs
- B In a locked cage at least 3 m from any oxygen cylinders
- C As close to the point of use as possible
- D Covered by a tarpaulin to shield the compressed cylinder from sunlight

**26.17** If you spill some oil on the floor and you do **not** have any absorbent material to clean the area, what should you do?

- A Spread it about to lessen the depth
- B Keep people out of the area and inform your supervisor
- C Do nothing, as it will eventually soak into the floor
- D Warn other people as they tread through it

**26.18** Where is the **safest** place to store refrigerant cylinders when they are **not** in use?

- A Outside in a special locked storage compound
- B In a company vehicle
- C Inside the building in a locked cupboard
- D In the immediate work area, ready for use the next day

26

**26.19** If refrigerant gases are released into a closed room in a building, what would they do?

A   Sink to the floor

B   Rise to the ceiling

C   Stay at the same level

D   Disperse safely within the room

**26.20** If you have to drill through a wall panel that you suspect contains an asbestos material, what should you do?

A   Ignore it and carry on

B   Put on safety goggles

C   Put on a dust mask

D   Stop work and report it

**26.21** When using a van to transport a refrigerant bottle, how should it be carried?

A   In the back of the van

B   In the passenger footwell of the van

C   In a purpose-built container within the rear of the van, with appropriate signage

D   In the van with all the windows open

**26.22** What safety devices should be fitted between the pipes and the gauges of oxyacetylene brazing equipment?

A   Non-return valves

B   On-off taps

C   Flame retardant tape

D   Flashback arresters

**26.23** Why is it essential to take great care when handling oxygen cylinders?

A   They contain highly flammable compressed gas

B   They contain highly flammable liquid gas

C   They are filled to extremely high pressures

D   They contain poisonous gas

**26.24** If you are working where welding is being carried out, what should be provided to protect you from welding flash?

A   A fire extinguisher

B   Warning notices

C   Screens

D   A hi-vis vest

26

**26.25** When using a blowtorch or brazing equipment to joint copper tube and fittings in a property, how should a fire extinguisher be made available?

A It should be available in the immediate work area

B It should be held over the joint while you are using the blowtorch

C It should be used to cool the fitting

D It should be available only if a property is occupied

**26.26** If you are carrying out hot works with a blowtorch, when should you stop using it?

A Just before you leave the site

B At least one hour before you leave the site

C At least two hours before you leave the site

D At least four hours before you leave the site

**26.27** Why must you never use oxygen when pressure testing?

A The molecules of the gas are too small

B The pressure required would not be reached

C When oxygen meets oil in a compressor it could explode and cause serious injury or death

D There is too much temperature pressure difference and a true record will not be given

**26.28** What should you ensure when pressure testing with nitrogen?

A That the nitrogen bottle is laid down to avoid it falling over

B That the nitrogen gauge can take the pressure required and that the bottle is secured upright

C That the temperature of the bottle is at room temperature to avoid a temperature pressure difference

D That purge brazing has taken place during the installation

**26.29** When planning a lifting operation how should the sequence of operations to enable a safe lift be confirmed?

A By verbal instruction

B In a method statement

C In a radio telephone message

D Via a notice in the canteen

**26.30** Which of these statements is true in relation to the safe working load (SWL) of lifting equipment?

A It is never marked on the equipment but kept with the test certificates

B It is provided for guidance only

C It may be exceeded by no more than 25%

D It is the absolute maximum safe working load

26

**26.31** What **must** be clearly marked on all lifting equipment?

A  The name of the manufacturer

B  The safe working load

C  The next test date

D  The specification of material from which it is made

**26.32** Which of these items of personal protective equipment (PPE) is designed to protect against infrared radiation damage to the eyes during flame cutting or welding?

A  Respiratory protective equipment (RPE)

B  Clear goggles

C  Eye protection with a tinted filter lens

D  Dust mask

**26.33** How should you position the bottles when using oxyacetylene brazing equipment?

A  Laid on their side and secured

B  Stood upright and secured

C  Stood upside down

D  Angled at 45°

**26.34** Which **two** of the following are essential safety checks that need to be carried out before using oxyacetylene equipment?

A  That the cylinders are full

B  That the cylinders, hoses and flashback arresters are in good condition

C  That the trolley wheels are the right size

D  That the area is well ventilated and clear of any obstructions

E  That the cylinders are the right weight

**26.35** When handling refrigerant gases, what personal protective equipment (PPE) should you wear as a **minimum**?

A  Eye protection, overalls, thermal-resistant gloves and helmet

B  Eye protection, overalls, thermal-resistant gloves and safety boots

C  Eye protection, overalls, harness and safety boots

D  Overalls, thermal-resistant gloves, helmet and safety boots

**26.36** What should you establish before entering a cold room?

A  The size of the cold room

B  The temperature of the cold room

C  Whether the exit door is fitted with an internal handle

D  Whether there are lights and power in the cold room

**26.37** What **must** you ensure before using a ladder?

- A) That it is secured to prevent it from moving sideways or sliding outwards
- B) That no-one else has booked the ladder for their work
- C) That an apprentice or workmate is standing by in case you slip and fall
- D) That the weather forecast is for a bright, clear day

**26.38** When positioning and erecting a stepladder, which of the following is essential for its safe use?

- A) It has a tool tray towards the top of the steps
- B) The restraint mechanism is spread to its full extent
- C) You will be able to reach the job by standing on the top step
- D) Your supervisor has positioned and erected the steps

**26.39** What is the recommended **maximum** height for a free-standing mobile tower?

- A) There is no restriction
- B) 2 m
- C) The height recommended by the manufacturer
- D) 12 m

**26.40** What is the **first** thing you should do after getting on to the platform of a correctly erected mobile tower?

- A) Check that the brakes are locked on
- B) Check for overhead power lines
- C) Close the access hatch to prevent falls of personnel, tools or equipment
- D) Make sure that the tower does not rock or wobble

**26.41** What should be done before a mobile tower is moved?

- A) All people and equipment must be removed from the platform
- B) A permit to work must be issued
- C) The principal contractor must give their approval
- D) Arrangements must be made with the forklift truck driver

**26.42** What must be done **first** before any roof work is carried out?

- A) A risk assessment must be carried out
- B) The operatives working on the roof must be trained in the use of safety harnesses
- C) Permits to work must be issued only to those allowed to work on the roof
- D) A weather forecast must be obtained

26

**26.43** What is edge protection designed to do?

A — Make access to the roof easier

B — Secure tools and materials close to the edge

C — Prevent rainwater running off the roof onto workers below

D — Prevent the fall of people and materials

---

**26.44** You have to carry out a job, over a few days, on the flat roof of a two-storey building, about 1 m from the edge of the roof, which has a low parapet. How should you reduce your risk of falling?

A — Carry on with the job, provided you don't get dizzy with heights

B — Use a full body harness, lanyard and anchor while doing the job

C — Ask for double guard-rails and toe-boards to be installed to prevent you falling

D — Get your mate to do the work, while you hold on to them

---

**26.45** You have been asked to install a number of ceiling-mounted air-conditioning units in a large, open-plan area, which has a good floor. What is the best way to access the work area?

A — From a stepladder

B — From scaffold boards and floor stands

C — By standing on packing cases

D — From a mobile tower

---

**26.46** When assembling a mobile tower what major hazard **must** you be aware of?

A — Water pipes

B — Cable trays

C — False ceilings

D — Overhead service cables

# 27    HVACR – Services and facilities maintenance

**27.01** When arriving at an occupied building, who or what should you consult to find out about any asbestos in the premises before starting work?

- A. The person responsible for the building, so you can view the asbestos register
- B. The building's receptionist
- C. The building's logbook
- D. The building's caretaker

**27.02** Where might you find information on the safe way to maintain the services in a building?

- A. The noticeboard
- B. The safety officer
- C. The local Health and Safety Executive (HSE) office
- D. The health and safety file for the building

**27.03** In the normal office environment what should be the hot water temperature at the tap furthest from the boiler, after it has been run for one minute?

- A. At least 15°C
- B. At least 35°C
- C. At least 50°C
- D. At least 100°C

**27.04** What should be the maximum temperature for a cold water supply, after it has been run for one minute?

- A. 10°C
- B. 20°C
- C. 35°C
- D. 50°C

**27.05** What is required if there is a cooling tower on site?

- A. A formal logbook
- B. A written scheme of examination
- C. Regular visits by the Local Authority Environmental Health officer
- D. Inspections by the water supplier

**27.06** Which two of the following are classed as pressure systems?

- A. Medium and high temperature hot water systems at or above 95°C
- B. Cold water systems
- C. Steam systems
- D. Office tea urns
- E. Domestic heating systems

**27.07** What is required before a pressure system can be operated?

A. A written scheme of examination

B. A hot-work permit

C. Operative certification

D. A minimum of two competent persons to operate the system

**27.08** Which of these statements is true in relation to the water used in cooling systems?

A. It should be replaced annually

B. It should be chemically treated

C. It should be chilled

D. It should be drinking water

**27.09** Which of the following should you not do when replacing the filters in an air-conditioning system?

A. Put the old filters in a dustbin

B. Follow a job-specific risk assessment and method statement

C. Wear appropriate overalls

D. Wear a respirator

**27.10** After servicing a gas boiler, what checks must you make by law?

A. Check for water leaks

B. Check for flueing, ventilation, gas rate and safe functioning

C. Check the pressure relief valve

D. Check the thermostat setting

**27.11** What must you do before adding inhibitor to a heating system?

A. Check for leaks on the system

B. Raise the system to working temperature

C. Read the COSHH assessment for the product

D. Bleed the heat emitters

**27.12** How should access to a riser be controlled?

A. By a site security operative

B. By those who are working in it

C. By the main contractor

D. By a permit to work system

**27.13** What is the correct action to take if natural gas is detected in an underground service duct?

A — No action, as it is not harmful

B — Evacuate the duct

C — Carry on working but do not use electrical equipment

D — Carry on working until the end of the shift

**27.14** What should you consider **first** when planning to work in a confined space?

A — Whether the job has been priced properly

B — Whether sufficient resource has been allocated

C — Whether the correct tools have been arranged

D — Whether the work could be done in another way to avoid the need to enter the confined space

**27.15** How should an adequate supply of breathable fresh air be provided in a confined space when breathing apparatus is **not** being worn?

A — An opening in the top of the confined space

B — Forced mechanical ventilation

C — Natural ventilation

D — An opening at the bottom of the confined space

**27.16** What are the **two** main safety considerations when using oxyacetylene equipment in a confined space?

A — The hoses may not be long enough

B — Unburnt oxygen may cause an oxygen-enriched atmosphere

C — The burner will be hard to light

D — Wearing the correct goggles

E — The risk of a flammable gas leak

**27.17** What precaution should be taken to protect against lighting failure in a confined space?

A — Remember where you got in

B — Ensure it is daylight when you do the work

C — Each operative should carry a torch

D — Secure a rope near the entrance and trail it behind you so that you can trace your way back

**27.18** While working on your own and tracing pipework in a building, you notice that the pipes enter a service duct. What should you do?

A — Go into the service duct and continue to trace the pipework

B — Ask someone in the building to act as your second person

C — Put on your personal protective equipment (PPE) and carry on with the job

D — Stop work until a risk assessment has been carried out

27

**27.19** Before working on electrically powered equipment, what is the procedure to make sure that the supply is dead before work starts?

A Switch off and remove the fuses

B Switch off and cut through the supply with insulated pliers

C Test the circuit, switch off and isolate the supply at the mains board

D Switch off, isolate the supply at the mains board, lock out and tag

**27.20** What would you use to find out whether a wall into which you are about to drill contains an electric supply?

A A neon screwdriver

B A cable tracer

C A multimeter

D A hammer and chisel

**27.21** Which type of power drill is most suitable for fixing a run of pipework outside in wet weather?

A Battery-powered drill

B Drill with 110 volt power supply

C Drill with 24 volt power supply

D Any mains voltage drill with a power breaker

**27.22** Why is temporary continuity bonding carried out before removing and replacing sections of metallic pipework?

A To provide a continuous earth for the pipework installation

B To prevent any chance of blowing a fuse

C To maintain the live supply to the electrical circuit

D To prevent any chance of corrosion to the pipework

**27.23** What is the procedure for ensuring that the electrical supply is dead before replacing an electric immersion heater?

A Switch off and disconnect the supply to the immersion heater

B Switch off and cut through the electric cable with insulated pliers

C Switch off and test the circuit

D Lock off the supply, isolate at the mains board, test the circuit and hang a warning sign

**27.24** What is used to reduce 230 volts to 110 volts on site?

A Residual current device (RCD)

B Transformer

C Circuit breaker

D Step-down generator

27

**27.25** What colour power outlet on a portable generator would supply 230 volts?

A Black

B Blue

C Red

D Yellow

**27.26** What action should you take if a natural gas leak is reported in a closed area?

A Ventilate the area and phone the gas emergency service

B Establish whether or not it is safe to enter

C Turn off the light

D Wear safety footwear

**27.27** Which of the following actions should you take if a refrigerant leak is reported in a closed area?

A Switch off the system, ventilate the area and test to establish if it is safe to enter

B Trace the leak and try to make a temporary repair

C Leave the system running until all of the gas has leaked out and then make a repair

D No action is required, as refrigerant gas is completely harmless

**27.28** How is legionella transmitted?

A By breathing contaminated airborne water droplets

B Through human contact

C Through contact with dirty clothes

D Via rat urine

**27.29** Which of the following is the most likely place to find legionella?

A In drinking water

B In hot water taps above 50°C

C In infrequently used shower heads

D In a river

**27.30** If you spill some oil on the floor and you do not have any absorbent material to clean the area, what should you do?

A Spread it about to lessen the depth

B Keep people out of the area and inform your supervisor

C Do nothing, as it will eventually soak into the floor

D Warn other people as they tread through it

27

**27.31** How should liquefied petroleum gas (LPG) cylinders be carried to and from premises in a van?

A  In the back of the van

B  In the passenger footwell of the van

C  In a purpose-built container within the rear of the van, with appropriate signage

D  In the van with all the windows open

---

**27.32** When removing some panelling, you see a section of cabling with the wires showing. What should you do?

A  Carry on with your work, trying your best to avoid the cables

B  Touch the cables to see if they are live, and if so refuse to carry out the work

C  Wrap the defective cable with approved electrical insulation tape

D  Only work when the cable has been isolated or repaired by a competent person

---

**27.33** When must a shaft or pit be securely covered or have double guard-rails and toe-boards installed?

A  At a fall height of 1 m

B  When there is a potential risk of anyone falling into it

C  At a fall height of 2.5 m

D  At a fall height of 3 m

---

**27.34** What is the ideal temperature for legionella to breed?

Between 75°C and 100°C

Between 45°C and 75°C

Between 20°C and 45°C

Below 20°C

---

**27.35** When assembling a mobile access tower what major hazard must you be aware of?

A  Water pipes

B  Cable trays

C  False ceilings

D  Overhead service cables

---

**27.36** What should you do if, when carrying out a particular task, the correct tool is not available?

A  Wait until you have the appropriate tool for the task

B  Borrow a tool from the building caretaker

C  Use the best tool available in the toolkit

D  Modify one of the tools you have

**27.37** Who should be informed if a legionella outbreak is suspected?

A. The Health and Safety Executive (HSE)

B. The police

C. A coroner

D. The nearest hospital

**27.38** When should the use of a permit to work be considered?

A. For all high risk work activities

B. For all equipment isolations

C. At the beginning of each shift

D. When there is enough time to complete the paperwork

**27.39** Who should fit a padlock and tag to an electrical lock-out guard?

A. Anyone working on the system

B. Only the person who fitted the lock-out guard

C. The senior engineer

D. The site supervisor

**27.40** If you arrive on site and find the mains isolator for a component is switched off, what should you do?

A. Switch it on and get on with your work

B. Switch it on and check the safety circuits to see if there is a fault

C. Contact the person in control of the premises

D. Ask people around the building and, if no-one responds, switch it on and get on with your work

**27.41** When carrying out solvent welding on plastic pipework, what particular safety measure **must** you apply?

A. The area must be well ventilated

B. The supervisor must be present

C. The area must be enclosed

D. It must be done in daylight

**27.42** Before starting work on a particular piece of equipment, who or what should you consult?

A. The machine brochure

B. The operation and maintenance manual for the equipment

C. The manufacturer's data plate

D. The store person

27

**27.43** Which two of the following should a person who is going to work alone carry out to ensure their safety?

A Register their presence with the site representative before starting work

B Ensure their timesheet is accurate and countersigned

C Make sure that somebody regularly checks that they are OK

D Notify the site manager of the details of the work

E Only work outside of normal working hours

**27.44** What must you ensure to prevent unauthorised access to an unoccupied plant or switchgear room?

A That the access door is locked

B That a sign is posted

C That the power supply is isolated

D That a person is posted to prevent access

**27.45** When positioning and erecting a stepladder, which of the following is essential for its safe use?

A It has a tool tray towards the top of the steps

B The restraint mechanism is spread to its full extent

C You will be able to reach the job by standing on the top step

D Your supervisor has positioned and erected the steps

**27.46** What is the first thing you should do after getting on to the platform of a correctly erected mobile tower?

A Check that the brakes are locked on

B Check for overhead power lines

C Close the access hatch to prevent falls of personnel, tools or equipment

D Make sure that the tower does not rock or wobble

27

# 28    Plumbing (JIB)

**28.01** What should you do when using a blowtorch near to flexible pipe lagging?

A  Remove just enough lagging to carry out the work

B  Remove the lagging at least 1 m either side of the work

C  Remove the lagging at least 3 m either side of the work

D  Wet the lagging but leave it in place

**28.02** What is the most likely risk of injury when cutting large diameter pipe?

A  Your fingers may become trapped between the cutting wheel and the pipe

B  You may cut yourself on the inside edge of the cut pipe

C  You may damage your muscles due to continued use

D  A piece of sharp metal could fly off and hit you

**28.03** If you have been handling sheet lead, what is the most likely way lead could get into your bloodstream?

A  By not using the correct respirator

B  By not washing your hands before eating

C  By not changing out of your work clothes

D  By not wearing safety goggles

**28.04** The legionella bacteria that cause legionnaires' disease are most likely to be found in which of the following?

A  A boiler operating at a temperature of 80°C

B  A shower hose outlet

C  A cold water storage cistern containing water at 10°C

D  A toilet pan

**28.05** How are legionella bacteria passed on to humans?

A  Through fine water droplets, such as sprays or mists

B  By drinking dirty water

C  Through contact with the skin

D  From other people when they sneeze

**28.06** Which item of personal protective equipment (PPE) is designed to protect against infrared radiation damage to the eyes during flame cutting or welding?

A  Impact-rated safety goggles

B  Respiratory protective equipment (RPE)

C  Reflective vest

D  Eye protection with a tinted or filter lens

28

**28.07** If you are drilling a hole, when do you need to wear eye protection?

A Only when drilling overhead

B Only when the drill bit exceeds 20 mm

C Always, whatever the circumstances

D Only when drilling through concrete

**28.08** What should you do when repairing a burst water main using pipe-freezing equipment to isolate the damaged section of pipe?

A Always work in pairs when using pipe-freezing equipment

B Never allow the freezing gas to come into direct contact with surface water

C Never use pipe-freezing equipment on plastic pipe

D Wear gloves to avoid direct contact with the skin

**28.09** You are drilling a 100 mm diameter hole for a flue pipe through a brick wall with a large hammer drill. Which combination of personal protective equipment (PPE) should you be supplied with?

A Gloves, breathing apparatus and ear defenders

B Ear defenders, respiratory protective equipment and eye protection

C Ear defenders, respiratory protective equipment and barrier cream

D Barrier cream, boots and respiratory protective equipment

**28.10** While working, you come across a hard, white, powdery material that could be asbestos. What should you do?

A While wearing respiratory protective equipment, remove the material and dispose of it safely

B Remove the material, putting it back after finishing the job

C Stop work immediately and tell your supervisor about the material

D Dampen the material down with water and remove it before carrying out the work

**28.11** Why is it important that operatives know the difference between propane and butane equipment?

A Propane equipment operates at higher pressure

B Propane equipment operates at lower pressure

C Propane equipment is cheaper

D Propane equipment can be used with smaller, easy-to-handle cylinders

**28.12** Which of the following statements is true?

A Both propane and butane are heavier than air

B Butane is heavier than air while propane is lighter than air

C Propane is heavier than air while butane is lighter than air

D Both propane and butane are lighter than air

**28.13** Apart from the cylinders used in gas-powered forklift trucks, why should liquefied petroleum gas (LPG) cylinders **never** be placed on their side during use?

A — It would give a faulty reading on the contents gauge, resulting in flashback

B — Air could be drawn into the cylinder, creating a dangerous mixture of gases

C — The liquid gas would be at too low a level to allow the torch to burn correctly

D — The liquid gas could be drawn from the cylinder, creating a safety hazard

**28.14** What is the preferred method of checking for leaks when assembling liquefied petroleum gas (LPG) equipment before use?

A — Test with a lighted match

B — Sniff the connections to detect the smell of gas

C — Listen to hear for escaping gas

D — Apply leak detection fluid to the connections

**28.15** The use of oxyacetylene equipment is **not** recommended for which of the following jointing methods?

A — Jointing copper pipe using hard soldering

B — Jointing copper tube using capillary soldered fittings

C — Jointing mild steel tube

D — Jointing sheet lead

**28.16** Which of the following makes it essential to take great care when handling and transporting oxygen cylinders?

A — They contain highly flammable compressed gas

B — They contain highly flammable liquid gas

C — They are filled to extremely high pressures

D — They contain poisonous gas

**28.17** Where is the **safest** place to store oxyacetylene gas-welding bottles when they are **not** in use?

A — Outside in a special storage compound

B — In company vehicles

C — Inside the building in a locked cupboard

D — In the immediate work area, ready for use the next day

**28.18** Where should a fire extinguisher be if you are using a blowtorch to joint copper tube and fittings in a domestic property?

A — Available in the immediate work area

B — In your vehicle, as long as the doors are locked

C — There is no need for a fire extinguisher

D — Available only if a property is occupied

28

**28.19** What should you do if you are using a blowtorch near to timber?

A | Carry out the work taking care not to catch the timber

B | Use a non-combustible mat and have a fire extinguisher ready

C | Wet the timber first and have a bucket of water handy

D | Point the flame away from the timber and have a bucket of sand ready to put out the fire

**28.20** What is the colour of an acetylene cylinder?

A | Orange

B | Black

C | Green

D | Maroon

**28.21** You are required to replace below-ground drainage pipework in an excavation, which is approximately 2.5 m deep, but the trench sides show signs of collapse. What should you do?

A | Get on with the work as quickly as possible

B | Refuse to do the work until the trench sides have been properly supported

C | Get a mate to help you so that they can pass the materials down to you

D | Ensure that you do the work with a rope around you so that you can be pulled out

**28.22** You are working in an occupied building and have taken up six lengths of 3 m floor boarding when you are called away to an urgent job. What should you do?

A | Leave the job as it is

B | Cordon off the work area before leaving the job

C | Permanently re-fix the floorboards and floor coverings

D | Tell other workers to be careful while you are away

**28.23** You are preparing to use an electric-powered threading machine. Which of the following statements should apply?

A | The power supply should be 24 volts and the machine fitted with a guard

B | The power supply should be 400 volts and the machine fitted with a guard

C | Ensure your clothing cannot get caught on rotating parts of the machine

D | Ear defenders should be available and should be in good condition

**28.24** When replacing an electrical immersion heater in a hot water storage cylinder, what should you do to make sure that the electrical supply is dead before starting plumbing work?

A | Switch off and disconnect the supply to the immersion heater

B | Switch off and cut through the electric cable with insulated pliers

C | Switch off and test the circuit

D | Switch off, isolate the supply at the mains board and test the circuit

28

**28.25** Why should you carry out temporary continuity bonding before removing and replacing sections of metallic pipework?

A To provide a continuous earth for the pipework installation

B To prevent any chance of blowing a fuse

C To maintain the live supply to the electrical circuit

D To prevent any chance of corrosion to the pipework

**28.26** If you are required to re-fix a section of external rainwater pipe using a power drill in wet weather conditions, which type of drill is most suitable?

A Battery-powered drill

B Drill with 110 volt power supply

C Drill with 240 volt power supply

D Any mains voltage drill with a power breaker

**28.27** What piece of equipment would you use to find out whether a section of solid wall that you are about to drill into contains electric cables?

A A neon screwdriver

B A cable tracer

C A multimeter

D A hammer and chisel

**28.28** When is it safe to transport workers to the workplace in the rear of a van?

A When the driver has a heavy goods vehicle licence

B When the van is fitted with temporary seating

C When the van is fitted with proper seating and seat belts

D When the driver is over 21 years of age

**28.29** Which is the safest method of transporting long lengths of copper pipe by van?

A Tying the pipes to the roof with copper wire

B Someone holding the pipes on the roof rack as you drive along

C Putting the pipes inside the van with the ends out of the passenger window

D Using a pipe rack fixed to the roof of the van

**28.30** If lifting a roll of Code 5 sheet lead, what is the first thing you should do?

A Weigh the roll of lead

B Have a trial lift to see how heavy it feels

C Assess the whole task

D Ask your workmate to give you a hand

28

**28.31** What should you do if you need to move a cast iron bath, but it is too heavy to lift by yourself?

A Inform your supervisor and ask for assistance

B Get a lifting accessory

C Give it another try

D Try and find someone to give you a hand

**28.32** How should you install a flue into an existing chimney?

A Insert the liner from roof level using a roof ladder

B Work in pairs and insert the liner from roof level, working off the roof

C Work in pairs and insert the liner from roof level, working from a chimney scaffold

D Break through the chimney in the loft area and insert the liner from there

**28.33** If you are asked to move a cast iron boiler some distance, what should you do?

A Get a workmate to carry it

B Drag it

C Roll it end-over-end

D Use a suitable trolley or other manual handling aid

**28.34** If working on a plumbing job where noise levels are rather high, who would you expect to carry out noise assessment?

A A fully qualified plumber

B Your supervisor

C The site engineer

D A competent person

**28.35** You are removing guttering from a large, single-storey, metal-framed and cladded building and the job is likely to take all day. What is the most appropriate type of access equipment you could use?

A A ladder

B A mobile access tower

C A putlog scaffold

D A trestle scaffold

**28.36** What should you do if you arrive at a job which involves using ladder access to the roof and you notice that the ladder has been painted?

A Only use the ladder if it is made of metal

B Only use the ladder if it is made of wood

C Only use the ladder if wearing rubber-soled boots to prevent slipping

D Not use the ladder, and report the matter to your supervisor

**28.37** How should bottles be positioned when using oxyacetylene welding equipment?

A Laid on their side and secured

B Stood upright and secured

C Stood upside down

D Angled at 45°

28

# Further information

Answer pages                                          232

Acknowledgements

## 01    General responsibilities

| | | | |
|---|---|---|---|
| 1.01 | D | 1.15 | D |
| 1.02 | B | 1.16 | B |
| 1.03 | C | 1.17 | A, E |
| 1.04 | D, E | 1.18 | A |
| 1.05 | D | 1.19 | D |
| 1.06 | C | 1.20 | B |
| 1.07 | B, D | 1.21 | A |
| 1.08 | A, E | 1.22 | B |
| 1.09 | A | 1.23 | D |
| 1.10 | A | 1.24 | D |
| 1.11 | A | 1.25 | A |
| 1.12 | C | 1.26 | D |
| 1.13 | B | 1.27 | D |
| 1.14 | A | | |

## 02    Accident reporting and recording

| | | | |
|---|---|---|---|
| 2.01 | C | 2.11 | A |
| 2.02 | B | 2.12 | A |
| 2.03 | D | 2.13 | C |
| 2.04 | B, E | 2.14 | B |
| 2.05 | A, D | 2.15 | B, C |
| 2.06 | A | 2.16 | A, B |
| 2.07 | B | 2.17 | C |
| 2.08 | A | 2.18 | D |
| 2.09 | B | 2.19 | C |
| 2.10 | B | | |

## 03 First aid and emergency procedures

| | | | |
|---|---|---|---|
| 3.01 | A | 3.11 | D |
| 3.02 | A, B | 3.12 | A |
| 3.03 | B, C | 3.13 | C |
| 3.04 | B, C | 3.14 | A |
| 3.05 | C | 3.15 | A |
| 3.06 | A, B | 3.16 | C |
| 3.07 | D | 3.17 | B |
| 3.08 | A | 3.18 | B |
| 3.09 | D | 3.19 | C |
| 3.10 | D | 3.20 | C |

## 04 Personal protective equipment

| | | | |
|---|---|---|---|
| 4.01 | D | 4.15 | D |
| 4.02 | D | 4.16 | D |
| 4.03 | B | | |
| 4.04 | B | | |
| 4.05 | D | | |
| 4.06 | A | 4.17 | A |
| 4.07 | A | 4.18 | C |
| 4.08 | C | 4.19 | D |
| 4.09 | B | 4.20 | B |
| 4.10 | C | 4.21 | D |
| 4.11 | A | 4.22 | Face |
| 4.12 | D | | |
| 4.13 | D | | |
| 4.14 | B | | |

## 05   Environmental awareness and waste control

| | | | | | |
|---|---|---|---|---|---|
| 5.01 | | D | 5.15 | | B |
| 5.02 | | B | 5.16 | | D, E |
| 5.03 | | A, B | 5.17 | | B, C |
| 5.04 | | C | 5.18 | | B |
| 5.05 | | A, C, E | 5.19 | | A |
| 5.06 | | A, B | 5.20 | | D |
| 5.07 | | B | 5.21 | | C, E |
| 5.08 | | D | 5.22 | | B |
| 5.09 | | D, E | 5.23 | | A |
| 5.10 | Broken bricks | Non-hazardous | 5.24 | | A |
| | Untreated timber off-cuts | Non-hazardous | 5.25 | | B |
| | Flourescent light tubes | Hazardous | 5.26 | | D |
| | Oil-based paint | Hazardous | 5.27 | | D |
| 5.11 | | D | 5.28 | | D |
| 5.12 | | B, D | 5.29 | | D |
| 5.13 | | D | 5.30 | | B |
| 5.14 | | C | | | |

## 06 Dust and fumes (Respiratory hazards)

| | | | |
|---|---|---|---|
| 6.01 | D | 6.26 | D |
| 6.02 | A, C | 6.27 | A |
| 6.03 | A, B | 6.28 | A |
| 6.04 | D | 6.29 | B |
| 6.05 | A | 6.30 | A, B |
| 6.06 | B | 6.31 | D |
| 6.07 | A | 6.32 | C |
| 6.08 | A | 6.33 | C |
| 6.09 | A | 6.34 | A, B |
| 6.10 | A | 6.35 | D, E |
| 6.11 | A | 6.36 | D, E |
| 6.12 | A | 6.37 | B |
| 6.13 | B | 6.38 | D |
| 6.14 | B | 6.39 | C |
| 6.15 | C | 6.40 | C |
| 6.16 | B | 6.41 | C |
| 6.17 | C | 6.42 | A, B |
| 6.18 | A | 6.43 | B |
| 6.19 | D | 6.44 | A |
| 6.20 | C | 6.45 | A |
| 6.21 | C | 6.46 | D |
| 6.22 | D | 6.47 | A, E |
| 6.23 | C | 6.48 | B |
| 6.24 | A | 6.49 | D |
| | | 6.50 | A |
| | | 6.51 | A, D |
| | | 6.52 | A |
| 6.25 | B | 6.53 | D |

## 07 Noise and vibration

| | | | | |
|---|---|---|---|---|
| 7.01 | C, D | 7.14 | 2 m  3 m  2 m  4 | |
| 7.02 | D | | | |
| 7.03 | C | | | |
| 7.04 | C | | | |
| 7.05 | C | 7.15 | D | |
| 7.06 | B | 7.16 | D | |
| 7.07 | C | 7.17 | B | |
| 7.08 | B, D, E | 7.18 | C | |
| 7.09 | B | 7.19 | D | |
| 7.10 | C | 7.20 | D | |
| 7.11 | C | 7.21 | C | |
| 7.12 | C | 7.22 | C | |
| 7.13 | A, B | 7.23 | A | |
| | | 7.24 | A, B | |

## 08 Health and welfare

| | | | |
|---|---|---|---|
| 8.01 | A | 8.13 | A |
| 8.02 | B | 8.14 | D |
| 8.03 | B | 8.15 | D |
| 8.04 | A | 8.16 | B |
| 8.05 | C | 8.17 | B |
| 8.06 | C | 8.18 | C |
| 8.07 | D | 8.19 | Sharps |
| 8.08 | A, D | | |
| 8.09 | D | | |
| 8.10 | D | | |
| 8.11 | C | 8.20 | B |
| 8.12 | C | 8.21 | D |

## 08 Health and welfare (continued)

| | | | | |
|---|---|---|---|---|
| 8.22 | D | | 8.46 | B |
| 8.23 | A | | 8.47 | D |
| 8.24 | C, E | | 8.48 | D |
| 8.25 | B | | 8.49 | D |
| 8.26 | B | | 8.50 | A |
| 8.27 | D | | 8.51 | D |
| 8.28 | D | | 8.52 | A |
| 8.29 | D | | 8.53 | A |
| 8.30 | B | | 8.54 | B |
| 8.31 | A | | 8.55 | C |
| 8.32 | B | | 8.56 | D |
| 8.33 | B | | 8.57 | A |
| 8.34 | A | | 8.58 | B |
| 8.35 | B | | 8.59 | B |
| 8.36 | D | | 8.60 | A |
| 8.37 | A | | 8.61 | D |
| 8.38 | A | | 8.62 | A |
| 8.39 | A | | 8.63 | C |
| 8.40 | B | | 8.64 | B |
| 8.41 | A | | 8.65 | D |
| 8.42 | C | | 8.66 | A |
| 8.43 | B | | 8.67 | B |
| 8.44 | A | | 8.68 | C |
| 8.45 | B | | | |

## 09 Manual handling

| | |
|---|---|
| 9.01 | B |
| 9.02 | C |
| 9.03 | B |
| 9.04 | B |
| 9.05 | B, E |
| 9.06 | A |
| 9.07 | C |
| 9.08 | B |
| 9.09 | C |
| 9.10 | A |

| | |
|---|---|
| 9.11 | Back |

| | |
|---|---|
| 9.12 | B |
| 9.13 | A, B, D |
| 9.14 | C, E |
| 9.15 | A, C, D |
| 9.16 | A |
| 9.17 | A |
| 9.18 | B |
| 9.19 | C |
| 9.20 | C |
| 9.21 | B |
| 9.22 | A |

# 10  Safety signs

| | | | |
|---|---|---|---|
| 10.01 | C | 10.03 | D |
| 10.02 | A, B, E | | |

# 11  Fire prevention and control

| | | | |
|---|---|---|---|
| 11.01 | C, E | 11.10 | C |
| 11.02 | D, E | 11.11 | B |
| 11.03 | C | 11.12 | B, C |
| 11.04 | A | 11.13 | 1-C, 2-A, 3-B, 4-D |
| 11.05 | B | 11.14 | A |
| 11.06 | A, B, D | 11.15 | A |
| 11.07 | A | 11.16 | A |
| 11.08 | D | 11.17 | B |
| 11.09 | C, D | 11.18 | B |

# 12  Electrical safety, tools and equipment

| | | | |
|---|---|---|---|
| 12.01 | B | 12.12 | B |
| 12.02 | D | 12.13 | D |
| 12.03 | D, E | 12.14 | D |
| 12.04 | D | 12.15 | D |
| 12.05 | C | 12.16 | B |
| 12.06 | B | 12.17 | B, E |
| 12.07 | B | 12.18 | B, D |
| 12.08 | A | 12.19 | A, D |
| 12.09 | D | 12.20 | B |
| 12.10 | D | 12.21 | C |
| 12.11 | D | | |

## 12    Electrical safety, tools and equipment (continued)

| | | | |
|---|---|---|---|
| 12.22 | | 12.28 | B |
| | | 12.29 | B |
| | | 12.30 | C |
| | | 12.31 | B, D, E |
| 12.23 | A | 12.32 | A, C, D |
| 12.24 | B | 12.33 | A |
| 12.25 | C | 12.34 | C, D |
| 12.26 | A | 12.35 | D |
| 12.27 | D | | |

## 13 Site transport and lifting operations

| | | | |
|---|---|---|---|
| 13.01 | C | 13.20 | C |
| 13.02 | C | 13.21 | D |
| 13.03 | A | 13.22 | B |
| 13.04 | B | 13.23 | D |
| 13.05 | A | 13.24 | D |
| 13.06 | D | 13.25 | B |
| 13.07 | C | 13.26 | A |
| 13.08 | C | 13.27 | C |
| 13.09 | B | 13.28 | A |
| 13.10 | B | 13.29 | B |
| 13.11 | C | 13.30 | B |
| 13.12 | A | 13.31 | D |
| 13.13 | A | 13.32 | C |
| 13.14 | B | 13.33 | A |
| 13.15 | A | 13.34 | A |
| 13.16 | B | 13.35 | C |
| 13.17 | C | 13.36 | B |
| 13.18 | A | 13.37 | B |
| 13.19 | A, D | | |

## 14 Working at height

| | | | |
|---|---|---|---|
| 14.01 | B | 14.07 | 75° |
| 14.02 | A | | |
| 14.03 | D | | |
| 14.04 | D | | |
| 14.05 | B | | |
| 14.06 | B | 14.08 | C |
| | | 14.09 | D |
| | | 14.10 | A |

75°  85  75°  80

## 14 Working at height (continued)

| | | | |
|---|---|---|---|
| 14.11 | D | 14.35 | A |
| 14.12 | A | 14.36 | B |
| 14.13 | A | 14.37 | D |
| 14.14 | D | 14.38 | B |
| 14.15 | A | 14.39 | A |
| 14.16 | A | 14.40 | D |
| 14.17 | C | 14.41 | D |
| 14.18 | B | 14.42 | C |
| 14.19 | B | 14.43 | A |
| 14.20 | B | 14.44 | B |
| 14.21 | B | 14.45 | B |
| 14.22 | B | 14.46 | C |
| 14.23 | D | 14.47 | D |
| 14.24 | C | 14.48 | D |
| 14.25 | C | 14.49 | B |
| 14.26 | A | 14.50 | A |
| 14.27 | A | 14.51 | C |
| 14.28 | C | 14.52 | C |
| 14.29 | D | 14.53 | C |
| 14.30 | C | 14.54 | B |
| 14.31 | B | 14.55 | A |
| 14.32 | B, D | 14.56 | A |
| 14.33 | A | 14.57 | A |
| 14.34 | A | 14.58 | B |

## 15     Excavations and confined spaces

| | | | | |
|---|---|---|---|---|
| 15.01 | A | | 15.11 | C |
| 15.02 | C | | 15.12 | D |
| 15.03 | D | | 15.13 | A, D |
| 15.04 | D | | 15.14 | D |
| 15.05 | B | | 15.15 | C |
| 15.06 | A | | 15.16 | A |
| 15.07 | D | | 15.17 | C |
| 15.08 | C | | 15.18 | D |
| 15.09 | B | | 15.19 | B |
| 15.10 | A, B, D | | | |

## 16     Hazardous substances

| | | | | |
|---|---|---|---|---|
| 16.01 | C | | 16.10 | D |
| 16.02 | A | | 16.11 | D |
| 16.03 | D | | 16.12 | C |
| 16.04 | B | | 16.13 | B |
| 16.05 | D | | 16.14 | D |
| 16.06 | D | | 16.15 | C |
| 16.07 | A | | 16.16 | A |
| 16.08 | D | | 16.17 | A |
| 16.09 | C | | 16.18 | D |

## 17 Supervisory

| | | | | |
|---|---|---|---|---|
| 17.01 | A | 17.29 | A |
| 17.02 | B | 17.30 | A |
| 17.03 | D | 17.31 | B |
| 17.04 | B | 17.32 | A |
| 17.05 | B | 17.33 | A |
| 17.06 | A | 17.34 | B |
| 17.07 | A, E | 17.35 | B |
| 17.08 | C | 17.36 | B |
| 17.09 | B, E | 17.37 | B |
| 17.10 | C | 17.38 | A |
| 17.11 | B | 17.39 | D |
| 17.12 | D | 17.40 | D |
| 17.13 | B | 17.41 | D |
| 17.14 | B | 17.42 | A |
| 17.15 | C | 17.43 | C |
| 17.16 | C | 17.44 | C |
| 17.17 | A | 17.45 | D |
| 17.18 | B | 17.46 | D |
| 17.19 | B | 17.47 | D |
| 17.20 | A | 17.48 | B |
| 17.21 | B | 17.49 | 470 mm |
| 17.22 | A, B | | |
| 17.23 | C | | |
| 17.24 | C | | |
| 17.25 | C | 17.50 | 950 mm |
| 17.26 | B | | |
| 17.27 | D | | |
| 17.28 | B | | |

17.49   470 mm

470 mm

17.50   950 mm

950 mm

## 17    Supervisory (continued)

| | | | | |
|---|---|---|---|---|
| 17.51 | B | 17.62 | 9 m | |

9 m

| | | |
|---|---|---|
| 17.63 | | C |
| 17.64 | | A, B, E |
| 17.65 | | D |
| 17.66 | | A |
| 17.67 | | D |
| 17.68 | | B |
| 17.69 | | C |
| 17.70 | | C |
| 17.71 | | A |
| 17.72 | | C |

| | |
|---|---|
| 17.51 | B |
| 17.52 | B |
| 17.53 | B |
| 17.54 | D |
| 17.55 | C |
| 17.56 | B |
| 17.57 | A |
| 17.58 | D |
| 17.59 | A |
| 17.60 | D |
| 17.61 | C |

## 18    Demolition

| | | | | |
|---|---|---|---|---|
| 18.01 | B | 18.10 | | C |
| 18.02 | C | 18.11 | | D |
| 18.03 | A | 18.12 | | A |
| 18.04 | B | 18.13 | | B |
| 18.05 | C | 18.14 | | B |
| 18.06 | B | 18.15 | | B |
| 18.07 | 3 m | 18.16 | | D |

3 m

| | | | |
|---|---|---|---|
| 18.17 | | | D |
| 18.18 | | | A |
| 18.19 | | | A |
| 18.08 | B | 18.20 | B |
| 18.09 | D | 18.21 | C |

## 18    Demolition (continued)

| | | | | |
|---|---|---|---|---|
| 18.22 | D | | 18.32 | C |
| 18.23 | B | | 18.33 | D |
| 18.24 | C | | 18.34 | D |
| 18.25 | C | | 18.35 | A |
| 18.26 | A | | 18.36 | D |
| 18.27 | B | | 18.37 | A |
| 18.28 | B, C | | 18.38 | D |
| 18.29 | A | | 18.39 | B |
| 18.30 | B | | 18.40 | C |
| 18.31 | C | | 18.41 | C |

## 19    Highway works

| | | | | |
|---|---|---|---|---|
| 19.01 | A | | 19.17 | B, E |
| 19.02 | B, D | | 19.18 | C |
| 19.03 | B, C | | 19.19 | C |
| 19.04 | D | | 19.20 | C |
| 19.05 | D | | 19.21 | A |
| 19.06 | B | | 19.22 | B |
| 19.07 | D | | 19.23 | C |
| 19.08 | B | | 19.24 | A, E |
| 19.09 | A | | 19.25 | B |
| 19.10 | D | | 19.26 | B |
| 19.11 | C | | 19.27 | C |
| 19.12 | D | | 19.28 | A |
| 19.13 | A | | 19.29 | C |
| 19.14 | B | | 19.30 | B |
| 19.15 | C | | 19.31 | B |
| 19.16 | B | | 19.32 | B, E |

## 19    Highway works (continued)

| | | | |
|---|---|---|---|
| 19.33 | C | 19.43 | B |
| 19.34 | A | 19.44 | D |
| 19.35 | C | 19.45 | B |
| 19.36 | D | 19.46 | D |
| 19.37 | C | 19.47 | D |
| 19.38 | A | 19.48 | B |
| 19.39 | B | 19.49 | D |
| 19.40 | B | 19.50 | D |
| 19.41 | C | 19.51 | A |
| 19.42 | C | 19.52 | B |

## 20    Specialist work at height

| | | | |
|---|---|---|---|
| 20.01 | A, C, E | 20.14 | C |
| 20.02 | B | 20.15 | A |
| 20.03 | C | 20.16 | C |
| 20.04 | C | 20.17 | C |
| 20.05 | A | 20.18 | C, D |
| 20.06 | C | 20.19 | B |
| 20.07 | C | 20.20 | D |
| 20.08 | D | 20.21 | D |
| 20.09 | D | 20.22 | B |
| 20.10 | 470 m | 20.23 | C |
| | | 20.24 | D |
| | | 20.25 | C |
| | | 20.26 | B |
| 20.11 | B | 20.27 | B |
| 20.12 | C | 20.28 | B |
| 20.13 | B | 20.29 | C |

470 mm

## 20 Specialist work at height (continued)

| | | | |
|---|---|---|---|
| 20.30 | D | 20.39 | D |
| 20.31 | A | 20.40 | B |
| 20.32 | C | 20.41 | A |
| 20.33 | C | 20.42 | A |
| 20.34 | D | 20.43 | B |
| 20.35 | D | 20.44 | C |
| 20.36 | B, D, E | 20.45 | A |
| 20.37 | D | 20.46 | C |
| 20.38 | C, E | 20.47 | C |

## 21 Lifts and escalators

| | | | |
|---|---|---|---|
| 21.01 | B | 21.18 | B |
| 21.02 | D | 21.19 | C |
| 21.03 | A | 21.20 | B |
| 21.04 | D | 21.21 | A |
| 21.05 | A | 21.22 | C |
| 21.06 | B | 21.23 | A, C |
| 21.07 | C | 21.24 | A |
| 21.08 | D | 21.25 | B |
| 21.09 | B | 21.26 | A |
| 21.10 | A | 21.27 | C |
| 21.11 | B | 21.28 | C |
| 21.12 | B | 21.29 | D |
| 21.13 | C | 21.30 | A |
| 21.14 | C | 21.31 | B |
| 21.15 | A | 21.32 | A |
| 21.16 | B | 21.33 | B |
| 21.17 | B | 21.34 | A |

## 21    Lifts and escalators (continued)

| | | | |
|---|---|---|---|
| 21.35 | A | 21.43 | B |
| 21.36 | C | 21.44 | D |
| 21.37 | B | 21.45 | B |
| 21.38 | A | 21.46 | A |
| 21.39 | C | 21.47 | D |
| 21.40 | B | 21.48 | A |
| 21.41 | A | 21.49 | A |
| 21.42 | C | | |

## 22    Tunnelling

| | | | |
|---|---|---|---|
| 22.01 | B | 22.19 | D |
| 22.02 | C | 22.20 | D |
| 22.03 | B | 22.21 | B |
| 22.04 | C, E | 22.22 | D |
| 22.05 | C | 22.23 | A |
| 22.06 | D | 22.24 | B |
| 22.07 | C | 22.25 | A |
| 22.08 | D | 22.26 | A, B |
| 22.09 | C | 22.27 | A, B |
| 22.10 | B | 22.28 | A |
| 22.11 | B, E | 22.29 | A |
| 22.12 | B | 22.30 | C |
| 22.13 | D | 22.31 | A |
| 22.14 | B | 22.32 | A |
| 22.15 | B | 22.33 | D |
| 22.16 | D | 22.34 | A |
| 22.17 | D | 22.35 | B, E |
| 22.18 | A | 22.36 | B |

## 22 Tunnelling (continued)

| | | | | |
|---|---|---|---|---|
| 22.37 | B | 22.41 | B |
| 22.38 | C | 22.42 | A |
| 22.39 | D | 22.43 | C |
| 22.40 | 1.2 m | 22.44 | D |

1.2 m

## 23 HVACR – Heating and plumbing services

| | | | |
|---|---|---|---|
| 23.01 | C | 23.20 | A |
| 23.02 | D | 23.21 | B |
| 23.03 | D | 23.22 | A |
| 23.04 | C | 23.23 | C |
| 23.05 | A | 23.24 | B |
| 23.06 | B | 23.25 | A |
| 23.07 | A | 23.26 | B |
| 23.08 | A | 23.27 | D |
| 23.09 | A | 23.28 | B |
| 23.10 | B | 23.29 | C |
| 23.11 | D | 23.30 | D |
| 23.12 | A | 23.31 | D |
| 23.13 | B | 23.32 | A |
| 23.14 | B | 23.33 | A |
| 23.15 | C | 23.34 | B |
| 23.16 | B | 23.35 | C |
| 23.17 | D | 23.36 | C |
| 23.18 | D | 23.37 | A |
| 23.19 | C | 23.38 | A |

## 23 HVACR – Heating and plumbing services (continued)

| | | | | |
|---|---|---|---|---|
| 23.39 | D | | 23.42 | B |
| 23.40 | D | | 23.43 | C |
| 23.41 | B | | 23.44 | B |

## 24 HVACR – Pipefitting and welding

| | | | | |
|---|---|---|---|---|
| 24.01 | C | | 24.24 | B |
| 24.02 | C | | 24.25 | B |
| 24.03 | B | | 24.26 | A |
| 24.04 | D | | 24.27 | D |
| 24.05 | D | | 24.28 | D |
| 24.06 | C | | 24.29 | C |
| 24.07 | B | | 24.30 | B |
| 24.08 | A | | 24.31 | B |
| 24.09 | A | | 24.32 | A |
| 24.10 | D | | 24.33 | C, D |
| 24.11 | A | | 24.34 | B |
| 24.12 | B | | 24.35 | D |
| 24.13 | B | | 24.36 | B |
| 24.14 | D | | 24.37 | B, D |
| 24.15 | A | | 24.38 | C |
| 24.16 | A | | 24.39 | C |
| 24.17 | D | | 24.40 | A |
| 24.18 | D | | 24.41 | B |
| 24.19 | B | | 24.42 | C |
| 24.20 | C | | 24.43 | A |
| 24.21 | C | | 24.44 | A |
| 24.22 | C | | 24.45 | D |
| 24.23 | C | | 24.46 | C |

## 24 HVACR – Pipefitting and welding (continued)

| | | | |
|---|---|---|---|
| 24.47 | D | 24.48 | B |

## 25 HVACR – Ductwork

| | | | |
|---|---|---|---|
| 25.01 | C | 25.24 | D |
| 25.02 | D | 25.25 | D |
| 25.03 | D | 25.26 | C |
| 25.04 | C | 25.27 | A |
| 25.05 | A | 25.28 | A |
| 25.06 | B | 25.29 | D |
| 25.07 | A | 25.30 | C |
| 25.08 | D | 25.31 | A, E |
| 25.09 | A | 25.32 | B |
| 25.10 | B | 25.33 | B |
| 25.11 | C | 25.34 | A |
| 25.12 | B | 25.35 | B |
| 25.13 | A | 25.36 | C |
| 25.14 | D | 25.37 | C |
| 25.15 | B, D | 25.38 | A |
| 25.16 | C | 25.39 | A |
| 25.17 | B | 25.40 | D |
| 25.18 | D | 25.41 | B |
| 25.19 | B | 25.42 | B |
| 25.20 | C | 25.43 | C |
| 25.21 | C | 25.44 | D |
| 25.22 | B | 25.45 | D |
| 25.23 | D | | |

## 26 HVACR – Refrigeration and air conditioning

| | | | |
|---|---|---|---|
| 26.01 | C | 26.24 | C |
| 26.02 | A | 26.25 | A |
| 26.03 | B | 26.26 | B |
| 26.04 | A | 26.27 | C |
| 26.05 | B | 26.28 | B |
| 26.06 | C | 26.29 | B |
| 26.07 | D | 26.30 | D |
| 26.08 | C | 26.31 | B |
| 26.09 | A | 26.32 | C |
| 26.10 | B | 26.33 | B |
| 26.11 | B | 26.34 | B, D |
| 26.12 | D | 26.35 | B |
| 26.13 | B | 26.36 | C |
| 26.14 | D | 26.37 | A |
| 26.15 | A | 26.38 | B |
| 26.16 | B | 26.39 | C |
| 26.17 | B | 26.40 | C |
| 26.18 | A | 26.41 | A |
| 26.19 | A | 26.42 | A |
| 26.20 | D | 26.43 | D |
| 26.21 | C | 26.44 | C |
| 26.22 | D | 26.45 | D |
| 26.23 | C | 26.46 | D |

## 27 HVACR – Services and facilities maintenance

| | | | |
|---|---|---|---|
| 27.01 | A | 27.04 | B |
| 27.02 | D | 27.05 | A |
| 27.03 | C | 27.06 | A, C |

## 27 HVACR – Services and facilities maintenance (continued)

| | | | |
|---|---|---|---|
| 27.07 | A | 27.29 | C |
| 27.08 | B | 27.30 | B |
| 27.09 | A | 27.31 | C |
| 27.10 | B | 27.32 | D |
| 27.11 | C | 27.33 | B |
| 27.12 | D | 27.34 | Between 20°C and 45°C |
| 27.13 | B | | |
| 27.14 | D | | |
| 27.15 | B | | |
| 27.16 | B, E | 27.35 | D |
| 27.17 | C | 27.36 | A |
| 27.18 | D | 27.37 | A |
| 27.19 | D | 27.38 | A |
| 27.20 | B | 27.39 | A |
| 27.21 | A | 27.40 | C |
| 27.22 | A | 27.41 | A |
| 27.23 | D | 27.42 | B |
| 27.24 | B | 27.43 | A, C |
| 27.25 | B | 27.44 | A |
| 27.26 | A | 27.45 | B |
| 27.27 | A | 27.46 | C |
| 27.28 | A | | |

## 28 Plumbing (JIB)

| | | | |
|---|---|---|---|
| 28.01 | B | 28.05 | A |
| 28.02 | B | 28.06 | D |
| 28.03 | B | 28.07 | C |
| 28.04 | B | 28.08 | D |

## 28 Plumbing (JIB) (continued)

| | | | | |
|---|---|---|---|---|
| 28.09 | B | | 28.24 | D |
| 28.10 | C | | 28.25 | A |
| 28.11 | A | | 28.26 | A |
| 28.12 | A | | 28.27 | B |
| 28.13 | D | | 28.28 | C |
| 28.14 | D | | 28.29 | D |
| 28.15 | B | | 28.30 | C |
| 28.16 | C | | 28.31 | A |
| 28.17 | A | | 28.32 | C |
| 28.18 | A | | 28.33 | D |
| 28.19 | B | | 28.34 | D |
| 28.20 | D | | 28.35 | B |
| 28.21 | B | | 28.36 | D |
| 28.22 | B | | 28.37 | B |
| 28.23 | C | | | |

# FURTHER INFORMATION

## Notes

**Notes**

# FURTHER INFORMATION

## Notes

## Notes

# FURTHER INFORMATION

## Notes

**Notes**

# FURTHER INFORMATION

## Notes